T0259016

# Linear Programming and Algorithms for Communication Networks

## A Practical Guide to Network Design, Control, and Management

# Linear Programming and Algorithms for Communication Networks

**A Practical Guide to Network Design, Control, and Management**

## Eiji Oki

CRC Press
Taylor & Francis Group
Boca Raton London New York

CRC Press is an imprint of the
Taylor & Francis Group, an **informa** business

# Contents

# Preface

The purpose of mathematical programming, or optimization, is to maximize or minimize an objective function considering some constraints. One of the applications of mathematical programming is to design and control communication networks, which consist of multitudes of nodes and links. For example, when the capacity of each link is given in a network, a key problem is to find an optimum set of routes on which a traffic flow from a source node to a destination node can be maximized. Another related example is as follows: when the capacity and cost of each link in a network and a traffic demand from a source node to a destination node are given, a frequent problem is to find an optimum set of routes that minimizes the total cost of transmitting the required traffic demand. These problems are solved using the techniques raised in the field of mathematical programming. Linear Programming (LP) is a special case of mathematical programming, where the objective function and all the constraints are expressed as linear functions. Because most of many basic and fundamental optimization problems on communication networks are categorized into LP problems, this book focuses on LP.

There are several excellent books that well describe LP and its applications to communication networks for undergraduate and graduate students. Most of them explain how to theoretically solve optimization problems, while those on communication networks may provide some simple examples of typical applications of LP to communication networks by formulating problems on network design and control.

When network operators or service providers design and control their networks in practical environments, in most cases they first formulate an optimization problem that corresponds to the desired communication networks with required parameters, and they solve the problem by running an LP solver on a computer. The engineers want to know how to apply LP to network design and control in their practical situations. However, there is a gap between the theory of LP in the literature and its practical implementation. This book was therefore written to fill this gap.

This book is intended to provide the fundamentals of LP as applied to communication networks and a practical guide on how to solve the communication-related problems using an LP solver. For this purpose, the GNU Linear Programming Kit (GLPK) package, which is intended for solving LP, integer linear programming (ILP), and mixed integer linear programming (MILP) problems, is adopted in this book. GLPK is freely available. This book introduces and explains typical practical problems for communication networks and their solutions by providing sufficient programs for GLPK. GLPK supports the GNU MathProg modeling language, which is a subset of AMPL (a modeling language for mathematical programming). The language is supported by most popular commercial mathematical programming solvers, for example, CPLEX®. Once readers understand how to solve LP problems for communication networks using the GLPK descriptions in this book, they will also able to easily apply their knowledge to other solvers. The book also provides practical algorithms for these problems by showing helpful examples with demonstrations.

I have been using a draft of this book as the text for graduate courses and seminars at The University of Electro-Communications, Tokyo Japan. The draft has been continually enhanced to reflect the students' feedback since 2008. These courses and seminars in which this material has been used have attracted both academic and industrial practitioners, as design and control for communication networks are among the key topics in the information and communication technology industry. I was engaged in designing and controlling networks with NTT Laboratories, and have a rich background in practical networking technologies as well as advanced research and development activities.

Because current books are not sufficient to bridge the gap between the theory of LP and its practice for communication networks, I believe that this book will serve as a useful addition to the literature. This book describes not only fundamental and theoretical aspects, but also provides a practical guide to the understanding of network control and design using mathematical programming and algorithms.

# Audience

This book is a good text for senior and graduate students in electrical engineering, computer engineering, and computer science. Using it, students will understand both fundamental and advanced technologies so that they can better position themselves when they graduate and look for jobs in the networking field. This book is also intended for telecommunication/networking professionals, R&D managers, software and hardware engineers, system engineers, who are currently active or anticipate future involvement in networking, as it allows them to design networks and network elements, or more comprehensively collaborate with network designers in order to satisfy to their customers' needs.

The minimum requirement to understand this book is a knowledge of linear algebra and computer logic. Some background in communication networks would be useful. All the concepts in this book are developed from intuitive basics, with further insight provided through examples of practical applications.

# Organization

The book is organized as follows.

- Chapter 1 clearly describes optimization problems for communication networks, including the shortest path problem, max flow problem, and minimum-cost flow problem.

- Chapter 2 provides the fundamentals of linear programming and integer linear programming as required to address several problems; it includes an overview of the basic theory, formulations, and solutions.

- Chapter 3 introduces the LP solver, GLPK. How to obtain, install, and use it are explained.

- Chapter 4 deals with basic problems for communication networks, which are presented in Chapter 1. LP formulations and solutions using GLPK, and typical algorithms by showing intuitive examples, are presented for reference. GLPK tutorials for these problems are provided in detail.

- Chapter 5 presents several problems on finding disjoint paths for reliable communications. First, the basic problem of finding a set of disjoint routes whose total cost is minimal, called the MIN-SUM problem, is considered. Several approaches, which include ILP, a disjoint shortest pair algorithm, and Suurballe's algorithm, are introduced to solve the problem. Second, the MIN-SUM problem in a network with shared risk link groups (SRLGs) and its solutions are presented. Third, the MIN-SUM problem in a multiple-cost network and its solutions are introduced.

- Chapter 6 describes the optimization problems in optical wavelength-routed networks. For a basic optical network, the wavelength assignment problem is considered. It is transferred into a graph coloring problem, which is formulated as an ILP problem. The largest-degree-first approach, which is a heuristic algorithm, is presented as a fast-computation approach. Second, wavelength assignment for an optical network with multi-carrier distribution is also considered.

- Chapter 7 describes several routing strategies to maximize the network utilization for various traffic-demand models. One useful approach to enhancing routing performance is to minimize the maximum link utilization rate, also called the network congestion ratio, of all network links. Minimizing the network congestion ratio leads to an increase in admissible traffic.

- Chapter 8 presents routing problems in Internet Protocol (IP) networks. A link-state-based routing protocol is widely used in IP networks, where all packets are transmitted over shortest paths as determined by weights associated with each link in the network. Determining the optimal routing through shortest-path routing means determining the optimal link weights. This chapter deals with the optimization problem of finding a set of link weights against network failures.

- Chapter 9 presents mathematical puzzles that can be tackled by integer linear programming (ILP). They are the Sudoku puzzle, a river crossing puzzle, and a lattice puzzle. The ILP formulations and solutions by GLPK are presented. For the river crossing puzzle, the shortest path approach is also introduced to solve the problem.

## Programs and input data listed in this book

The programs and input data listed in this book are available at the following site: http://www.crcpress.com/.

## Acknowledgments

This book could not have been published without the help of many people. I thank them for their efforts in improving the quality of the book. I have done my best to accurately described linear programming and algorithms for communication networks as well as the basic concepts. I alone am responsible for any remaining error. If any error is found, please send an e-mail to eiji.oki@uec.ac.jp. I will correct them in future editions.

Several chapters of the book are based on our research work. I would like to thank the people who have contributed materials to some chapters. Especially, I thank Mohammad Kamrul Islam, Dr. Nattapong Kitsuwan, Ruchaneeya Leepila, and Dwina Fitriyandini Siswanto. I thank Yutaka Arai and Ihsen Aziz Ouédraogo for providing some numerical examples. The manuscript draft was reviewed by Seydou Ba and Abu Hena Al Muktadir. I am immensely grateful for their comments and suggestions.

I wish to thank my wife Naoko, my daughter Kanako, and my son Shunji, for their love and support.

Eiji Oki

## About the author

**Eiji Oki** is an associate professor at the University of Electro-Communications, Tokyo, Japan. He received B.E. and M.E. degrees in in-

strumentation engineering and a Ph.D. degree in electrical engineering from
Keio University, Yokohama, Japan, in 1991, 1993, and 1999, respectively. In
1993, he joined Nippon Telegraph and Telephone Corporation (NTT) Communication Switching Laboratories, Tokyo, Japan. He has been researching
network design and control, traffic-control methods, and high-speed switching
systems. From 2000 to 2001, he was a Visiting Scholar at the Polytechnic In-
stitute of New York University, Brooklyn, New York, where he was involved
in designing terabit switch/router systems. He was engaged in researching
and developing high-speed optical IP backbone networks with NTT Laboratories. He joined the University of Electro-Communications, Tokyo, Japan, in
July 2008. He has been active in standardization of path computation element
(PCE) and GMPLS in the IETF. He wrote more than ten IETF RFCs. He
served as a Guest Co-Editor for the Special Issue on "Multi-Domain Optical Networks: Issues and Challenges," June 2008, in *IEEE Communications
Magazine*; a Guest Co-Editor for the Special Issue on Routing, "Path Computation and Traffic Engineering in Future Internet," December 2007, in the
*Journal of Communications and Networks*; a Guest Co-Editor for the Special
Section on "Photonic Network Technologies in Terabit Network Era," April
2011, in *IEICE Transactions on Communications*; a Technical Program Com-
mittee (TPC) Co-Chair for the Conference on High-Performance Switching
and Routing in 2006, 2010, and 2012; a Track Co-Chair on Optical Networking for ICCCN 2009; a TPC Co-Chair for the International Conference on
IP+Optical Network (iPOP 2010 and 2012); and a Co-Chair of Optical Networks and Systems Symposium for IEEE ICC 2011. Professor Oki was the
recipient of the 1998 Switching System Research Award and the 1999 Excellent Paper Award presented by IEICE, the 2001 Asia-Pacific Outstanding
Young Researcher Award presented by IEEE Communications Society for
his contribution to broadband network, ATM, and optical IP technologies,
and the 2010 Telecom System Technology Prize by the Telecommunications
Advanced Foundation. He has co-authored three books, *Broadband Packet
Switching Technologies*, published by John Wiley, New York, in 2001, *GM-
PLS Technologies*, published by CRC Press, Boca Raton, Florida, in 2005,
and *Advanced Internet Protocols, Services, and Applications*, published by
Wiley in 2012. He is an IEEE Senior Member.

# Chapter 1

# Optimization problems for communications networks

Communication networks consist of nodes and links. Figure 1.1 shows an example of a network. This network consists of six nodes, node 1 to node 6. An arrow between two nodes is a connection, called a link, of those nodes. The traffic has a direction from the tail to the head of the arrow. For example, the arrow from node 1 to node 2 means that node 1 and node 2 are connected and the traffic flows from node 1 to node 2. The network in which each link has a direction, represented by a corresponding arrow, as shown in Figure 1.1, is called a directed graph. A number on each link indicates its link cost. In the case that the connection is represented by just a line, instead of an arrow, the traffic can flow in both directions on the link. A network with links through which the traffic flows in both directions is called a undirected graph.

This chapter introduces typical examples of the problems posed by communication networks, starting with the shortest path problem.

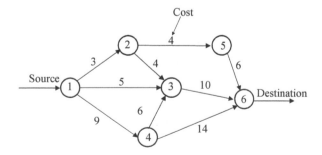

Figure 1.1: Network model with link costs.

1

## 1.1 Shortest path problem

Consider that node 1 wants transmit traffic to node 6, as shown in Figure 1.1. We need to find the path with the minimum cost to transmit the traffic. Nodes 1 and 6 are called source and destination nodes, respectively. The path with the minimum cost from the source node to the destination node is called the shortest path. The shortest path is determined by considering the link costs in the network. This problem is called the shortest path problem. The problem is solved and the solution is obtained, as shown in Figure 1.2. The shortest path from node 1 to node 6 is $1 \to 2 \to 5 \to 6$, and the path cost, which is the sum of costs of the links on the path, is $3 + 4 + 6 = 13$.

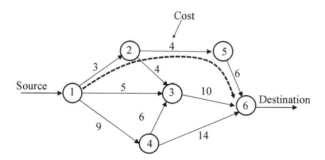

Figure 1.2: Shortest path from node 1 to node 6.

## 1.2 Max flow problem

Figure 1.3 shows a network that considers the capacity of each link. The number on each link represents the link capacity; that is the maximum traffic that can be transmitted through the link. Traffic volume, $v$, is injected from node 1. How much maximum traffic can we send from node 1 to node 6? Which route should the traffic be transmitted on? This problem is called the max flow problem. Figure 1.4 shows the solution of this problem. The maximum traffic volume from node 1 to node 6 is $v = 195$ and consists of five paths with their corresponding traffic volumes of $v_1$ to $v_5$. $v_1 = 15$ is sent on the first path, $1 \to 2 \to 5 \to 6$. $v_2 = 10$ is sent on the second path, $1 \to 2 \to 3 \to 6$. $v_3 = 100$ is sent on the third path, $1 \to 3 \to 6$. $v_4 = 60$ is sent on the fourth path, $1 \to 4 \to 3 \to 6$. $v_5 = 10$ is sent on the fifth path, $1 \to 4 \to 6$. The total traffic $v$ is $v_1 + v_2 + v_3 + v_4 + v_5 = 15 + 10 + 100 + 60 + 10 = 195$. The traffic that flows on each link does not exceed the link capacity. For example, the traffic on link $1 \to 2$ is $v_1 + v_2 = 15 + 10 = 25$, which does not exceed 25 (25 is the capacity of link $1 \to 2$). The traffic on link $3 \to 6$ is $v_2 + v_3 + v_4 = 10 + 100 + 60 = 170$ does not exceed 200 (200 is the capacity of link $1 \to 2$).

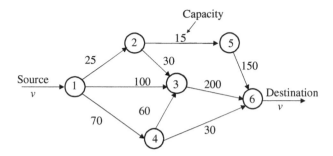

Figure 1.3: Network model with link capacities.

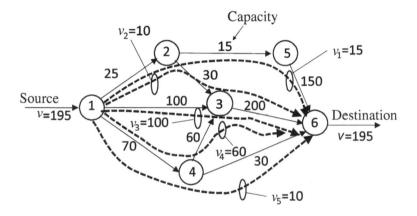

Figure 1.4: Max flow routing from node 1 to node 6.

## 1.3 Minimum-cost flow problem

Figure 1.5 shows a network that considers the cost and capacity of each link. The numbers on each link represents the link cost and the link capacity. The traffic flow cannot exceed the link capacity. The traffic volume that is required to be transmitted from a source node, node 1, to a destination node, node 6, is set to $v = 180$. How can we send the required traffic volume from node 1 to node 6 at the minimum cost? This problem is called the minimum-cost flow problem. In the minimum-cost flow, the required cost for each link is defined as the cost of the link × the traffic volume that flows on the link. We minimize the sum of costs on the path(s) to send the traffic from node 1 to node 6.

Figure 1.6 shows the solution of the minimum-cost flow problem. The traffic with the volume of $v$ is divided into five paths, from $v_1$ to $v_5$. $v_1 = 15$ is sent on the first path, $1 \rightarrow 2 \rightarrow 5 \rightarrow 6$. $v_2 = 10$ is sent on the second path, $1 \rightarrow 2 \rightarrow 3 \rightarrow 6$. $v_3 = 100$ is sent on the third path, $1 \rightarrow 3 \rightarrow 6$. $v_4 = 25$ is sent on the fourth path, $1 \rightarrow 4 \rightarrow 3 \rightarrow 6$. $v_5 = 30$ is sent on the fifth

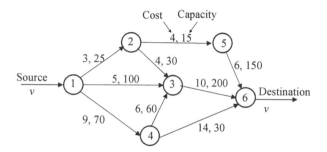

Figure 1.5: Network with link costs and capacities.

path, $1 \to 4 \to 6$. The total traffic volume is $v = v_1 + v_2 + v_3 + v_4 + v_5 = 15 + 10 + 100 + 25 + 30 = 180$. The total cost is 3180. There is no traffic flow that exceeds the capacity of the link on which it flows.

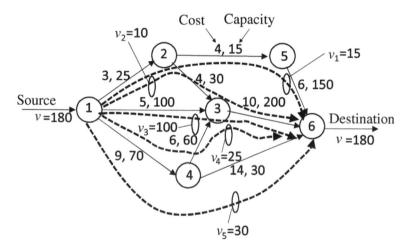

Figure 1.6: Minimum-cost flow from node 1 to node 6.

# Chapter 2

# Basics of linear programming

An optimization problem is a problem that aims to find the best solution from all feasible solutions. The best solution can be the minimum or maximum solution. An example of the former is finding the route from point A to point B that takes the shortest time. An example of the latter is determining how a production factory can maximize its profit using limited materials. Both problems are optimization problems. An optimization problem can be solved by mathematical programming, a technique that expresses and solves problems as mathematic models.

This chapter explains linear programming, which is a special case of mathematical programming.

## 2.1   Optimization problem

A businessman must travel from city A to city B on a business trip. He has two choices as to the means of transportation: airplane or train. How can he travel with the minimum cost given the following conditions?

- Condition 1: The price for a one-way ticket should not exceed $150.

- Condition 2: He should arrive at city B by 11:10 a.m.

- Condition 3: He should depart city A after 8:00 a.m.

He checks the airplane and train schedules, as listed in Table 2.1. There are eight choices. He has to choose one of them. He has to choose one out of eight choices; the one that satisfies all conditions and has the minimum cost. As all the prices in the table are less then $150, they satisfy condition 1. As for condition 2, choices 3 and 8 are cut because they arrive after 11:10 a.m. For condition 3, choices 1 and 4 are cut because their departure times are before

Table 2.1: Transportation details.

| Choice | Transportation | Departure time | Arrival time | Price ($) |
|--------|----------------|----------------|--------------|-----------|
| 1 | Airplane | 7:25 a.m. | 8:40 a.m. | 134.70 |
| 2 | Airplane | 9:50 a.m. | 11:05 a.m. | 136.70 |
| 3 | Airplane | 10:45 a.m. | 12:00 a.m. | 136.70 |
| 4 | Train | 7:56 a.m. | 10:36 a.m. | 138.50 |
| 5 | Train | 8:03 a.m. | 11:03 a.m. | 135.50 |
| 6 | Train | 8:20 a.m. | 10:56 a.m. | 138.50 |
| 7 | Train | 8:30 a.m. | 11:06 a.m. | 138.50 |
| 8 | Train | 8:33 a.m. | 11:30 a.m. | 135.50 |

8:00 a.m. Here, the businessman is left with choices 2, 5, 6, and 7. He refines the selection using the minimum cost, which is choice 5. Therefore, he will travel by train, leaving from city A at 8:03 a.m., and arriving at city B at 11:03 a.m., and spending $135.50.

An optimization problem consists of three components: decision variables, objective function, and constraints. In case of the above example, the decision variables are transportation, departure time, arrival time, and price. The objective function is the price. The constraints are conditions 1, 2, and 3. A mathematical model can be established that encompasses all three components.

- Decision variables: are the variables within a model that can be controlled. If there are $n$ decision variables, they are represented as $x_1, x_2, \cdots, x_n$.

- Objective function: is the function that we want to maximize or minimize. An objective function is written as $f(x_1, x_2, \cdots, x_n)$. If we want to maximize this function, we write

$$\max_{x_1, x_2, \cdots, x_n} f(x_1, x_2, \cdots, x_n). \tag{2.1}$$

If the function should be minimized, we express it by

$$\min_{x_1, x_2, \cdots, x_n} f(x_1, x_2, \cdots, x_n). \tag{2.2}$$

- Constraints: are conditions or limitations of the problem. Each is expressed in mathematical form as follows.

$$S_1(x_1, x_2, \cdots) \leq 0$$
$$S_2(x_1, x_2, \cdots) \leq 0$$
$$S_3(x_1, x_2, \cdots) \leq 0$$
$$\cdots \tag{2.3}$$

## 2.2 Linear programming problem

A linear programming (LP) problem is an optimization problem in which the objective function and all the constraints are expressed as linear functions. Even if just one of them is not a linear function, this problem is not an LP problem A linear function is expressed by

$$f(x_1, x_2, \dots) = a_1 x_1 + a_2 x_2 + \cdots + a_0, \tag{2.4}$$

where $a_1, a_2, \cdots, a_0$ are constants.

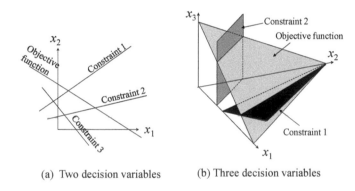

(a) Two decision variables     (b) Three decision variables

Figure 2.1: Linear programming problem.

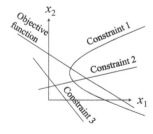

Figure 2.2: Example of nonlinear programming problem.

Figure 2.1 shows the appearance of linear functions. In Figure 2.1(a), there are two decision variables. The objective function and constraints are depicted by lines. In Figure 2.1(b), there are three decision variables. The objective function and decision variables are depicted by the planes. Figure 2.2 shows an example of a nonlinear programming (NLP) problem; obviously it is not an LP problem. The objective function and two constraints are linear functions, but one constraint is not a linear function. Therefore, this problem is not an LP problem.

Eqs. (2.5a)–(2.5f) show an LP problem. It consists of an objective function, constraints, and two decision variables, which are expressed by $x_1$ and $x_2$.

$$\text{Objective} \quad \max \quad x_1 + x_2 \tag{2.5a}$$
$$\text{Constraints} \quad 5x_1 + 3x_2 \leq 15 \tag{2.5b}$$
$$x_1 - x_2 \leq 2 \tag{2.5c}$$
$$x_2 \leq 3 \tag{2.5d}$$
$$x_1 \geq 0 \tag{2.5e}$$
$$x_2 \geq 0 \tag{2.5f}$$

In general, an LP problem that maximizes an objective function is represented by the following formula:

$$\text{Objective} \quad \max \quad c_1 x_1 + c_2 x_2 + \cdots + c_n x_n \tag{2.6a}$$
$$\text{Constraints} \quad a_{11} x_1 + a_{12} x_2 + \cdots + a_{1n} x_n \leq b_1 \tag{2.6b}$$
$$a_{21} x_1 + a_{22} x_2 + \cdots + a_{2n} x_n \leq b_2 \tag{2.6c}$$
$$\cdots$$
$$a_{m1} x_1 + a_{m2} x_2 + \cdots + a_{mn} x_n \leq b_m \tag{2.6d}$$
$$x_1 \geq 0 \tag{2.6e}$$
$$x_2 \geq 0 \tag{2.6f}$$
$$\cdots$$
$$x_n \geq 0 \tag{2.6g}$$

Eqs. (2.6e)–(2.6g) provide the ranges of the decision variables. Usually, Eqs. (2.6e)–(2.6g) are not necessary for the LP problem. However, their inclusion makes it easy to handle the LP problem in a consistent manner. Eqs. (2.6a)–(2.6g) are called a canonical form of an LP problem with maximization. They are also formulated by a matrix expression as follows:

$$\text{Objective} \quad \max \quad c^T x \tag{2.7a}$$
$$\text{Constraints} \quad Ax \leq b \tag{2.7b}$$
$$x \geq 0, \tag{2.7c}$$

where

$$\boldsymbol{x}^T = [x_1, \ldots, x_n] \tag{2.8a}$$

$$\boldsymbol{b}^T = [b_1, \ldots, b_m] \tag{2.8b}$$

$$\boldsymbol{c}^T = [c_1, \ldots, c_n] \tag{2.8c}$$

$$\boldsymbol{A} = \begin{bmatrix} a_{11} & a_{12} & \cdots & a_{1n} \\ a_{21} & a_{22} & \cdots & a_{2n} \\ \vdots & \vdots & \ddots & \vdots \\ a_{m1} & a_{m2} & \cdots & a_{mn} \end{bmatrix}. \tag{2.8d}$$

While Eqs. (2.5a)–(2.5f) represent an LP problem that maximizes an objective function, we can convert the objective function into a minimization problem. To maximize $x_1 + x_2$ is to minimize $-x_1 - x_2$. If we multiply the inequalities (2.5a)–(2.5d) by $-1$, the LP problem is transformed into

$$\begin{aligned} \text{Objective} \quad & \min \quad -x_1 - x_2 & (2.9a) \\ \text{Constraints} \quad & -5x_1 - 3x_2 \geq -15 & (2.9b) \\ & -x_1 + x_2 \geq -2 & (2.9c) \\ & -x_2 \geq -3 & (2.9d) \\ & x_1 \geq 0 & (2.9e) \\ & x_2 \geq 0. & (2.9f) \end{aligned}$$

An LP problem that minimizes an objective function is represented by the following formula:

$$\begin{aligned} \text{Objective} \quad & \min \quad c_1 x_1 + c_2 x_2 + \cdots + c_n x_n & (2.10a) \\ \text{Constraints} \quad & a_{11} x_1 + a_{12} x_2 + \cdots + a_{1n} x_n \geq b_1 & (2.10b) \\ & a_{21} x_1 + a_{22} x_2 + \cdots + a_{2n} x_n \geq b_2 & (2.10c) \\ & \cdots \\ & a_{m1} x_1 + a_{m2} x_2 + \cdots + a_{mn} x_n \geq b_m & (2.10d) \\ & x_1 \geq 0 & (2.10e) \\ & x_2 \geq 0 & (2.10f) \\ & \cdots \\ & x_n \geq 0. & (2.10g) \end{aligned}$$

Eqs. (2.10a)–(2.10g) are called a canonical form of an LP problem with minimization. They are also formulated by a matrix expression as follows:

$$\begin{aligned} \text{Objective} \quad & \min \quad \boldsymbol{c}^T \boldsymbol{x} & (2.11a) \\ \text{Constraints} \quad & \boldsymbol{A}\boldsymbol{x} \geq \boldsymbol{b} & (2.11b) \\ & \boldsymbol{x} \geq 0, & (2.11c) \end{aligned}$$

where

$$\boldsymbol{x}^T = [x_1, \ldots, x_n] \tag{2.12a}$$
$$\boldsymbol{b}^T = [b_1, \ldots, b_m] \tag{2.12b}$$
$$\boldsymbol{c}^T = [c_1, \ldots, c_n] \tag{2.12c}$$

$$\boldsymbol{A} = \begin{bmatrix} a_{11} & a_{12} & \cdots & a_{1n} \\ a_{21} & a_{22} & \cdots & a_{2n} \\ \vdots & \vdots & \ddots & \vdots \\ a_{m1} & a_{m2} & \cdots & a_{mn} \end{bmatrix}. \tag{2.12d}$$

Let us reconsider the LP problem of Eqs. (2.5a)–(2.5f). Let $y_1$, where $y_1 \geq 0$, be added to constraint $5x_1 + 3x_2 \leq 15$, expressed by Eq. (2.5b). We rewrite it as $5x_1 + 3x_2 + y_1 = 15$. Let $y_2$, where $y_2 \geq 0$, be added to constraint $x_1 - x_2 \leq 2$, expressed by Eq. (2.5c), and rewrite it as $x_1 - x_2 + y_2 = 2$. Let $y_3$, where $y_3 \geq 0$, be added to constraint $x_2 \leq 3$ expressed by Eq. (2.5d), and rewrite it as $x_2 + y_3 = 3$. A new LP problem is obtained by

$$\text{Objective} \quad \max \quad x_1 + x_2 \tag{2.13a}$$
$$\text{Constraints} \quad 5x_1 + 3x_2 + y_1 = 15 \tag{2.13b}$$
$$x_1 - x_2 + y_2 = 2 \tag{2.13c}$$
$$x_2 + y_3 = 3 \tag{2.13d}$$
$$x_1 \geq 0 \tag{2.13e}$$
$$x_2 \geq 0 \tag{2.13f}$$
$$y_1 \geq 0 \tag{2.13g}$$
$$y_2 \geq 0 \tag{2.13h}$$
$$y_3 \geq 0, \tag{2.13i}$$

where $y_1$, $y_2$, and $y_3$ are called slack variables.

In general, by introducing slack variables, an LP problem can be expressed in the following form:

$$\text{Objective} \quad \max \text{ or } \min \quad c_1 x_1 + c_2 x_2 + \cdots + c_n x_n \tag{2.14a}$$
$$\text{Constraints} \quad a_{11} x_1 + a_{12} x_2 + \cdots + a_{1n} x_n = b_1 \tag{2.14b}$$
$$a_{21} x_1 + a_{22} x_2 + \cdots + a_{2n} x_n = b_2 \tag{2.14c}$$
$$\cdots$$
$$a_{m1} x_1 + a_{m2} x_2 + \cdots + a_{mn} x_n = b_m \tag{2.14d}$$
$$x_1 \geq 0 \tag{2.14e}$$
$$x_2 \geq 0 \tag{2.14f}$$
$$\cdots$$
$$x_n \geq 0. \tag{2.14g}$$

Eqs. (2.14a)–(2.14g) are called a standard form of an LP problem. They are also formulated by a matrix expression as follows:

$$\text{Objective} \quad \max \text{or} \min \quad \boldsymbol{c}^T \boldsymbol{x} \tag{2.15a}$$

$$\text{Constraints} \quad \boldsymbol{A}\boldsymbol{x} = \boldsymbol{b} \tag{2.15b}$$

$$\boldsymbol{x} \geq 0, \tag{2.15c}$$

where

$$\boldsymbol{x}^T = [x_1, \ldots, x_n] \tag{2.16a}$$

$$\boldsymbol{b}^T = [b_1, \ldots, b_m] \tag{2.16b}$$

$$\boldsymbol{c}^T = [c_1, \ldots, c_n] \tag{2.16c}$$

$$\boldsymbol{A} = \begin{bmatrix} a_{11} & a_{12} & \cdots & a_{1n} \\ a_{21} & a_{22} & \cdots & a_{2n} \\ \vdots & \vdots & \ddots & \vdots \\ a_{m1} & a_{m2} & \cdots & a_{mn} \end{bmatrix}. \tag{2.16d}$$

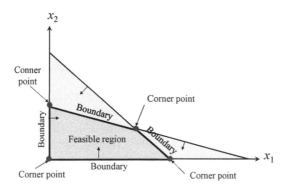

Figure 2.3: Nomenclature in linear programming problem.

Figure 2.3 shows terminology for an LP problem with two decision variables. A boundary is a constraint that expresses the upper or lower bound of an inequality or equality. The feasible region is an area delineated by the boundaries. A corner point is an intersection of the boundaries.

The following problem is an example of an LP problem with two decision variables, $x$ and $y$:

$$\text{Objective} \quad \max \quad x + y \tag{2.17a}$$
$$\text{Constraints} \quad 5x + 3y \le 15 \tag{2.17b}$$
$$x - y \le 2 \tag{2.17c}$$
$$y \le 3 \tag{2.17d}$$
$$x \ge 0 \tag{2.17e}$$
$$y \ge 0. \tag{2.17f}$$

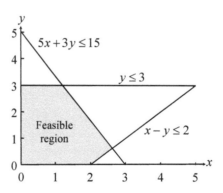

Figure 2.4: Constraints in linear programming problem.

Figure 2.4 shows ranges of constraints as Eqs. (2.17b)–(2.17f). The area bounded by the constraints is the feasible region. Let $z$ be the objective function, $z = x + y$. We want to maximize the objective function, $z$. We rewrite this function as $y = -x + z$. The slope of this function is $-1$, and this function intersects the $y$-axis at $(0, z)$. If we move this function up along the $y$-axis while keeping its slope, $z$ increases. On the other hand, if we move it down along the $y$-axis, $z$ decreases. As shown in Figure 2.5, the maximum value of $z$ is determined by moving the objective function up along the $y$-axis while keeping the slope, $-1$, under the condition that the function passes through the feasible region.

As shown in Figure 2.5(a), let us start from $y = -x + 0$. We then move the objective function up along the $y$-axis until it touches the corner point $(2, 0)$, as shown in Figure 2.5(b). The function becomes $y = -x + 2$. Next, move it up to touch the corner point $(0, 3)$, as shown in Figure 2.5(c). The function becomes $y = -x + 3$. We continue moving this function up until it reaches the end corner point of the feasible region, $(\frac{6}{5}, 3)$, as shown in Figure 2.5(d). The function becomes $y = -x + \frac{21}{5}$. The function is no longer moved up,

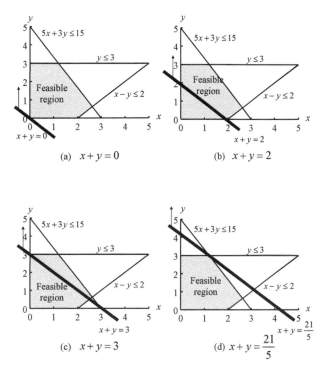

Figure 2.5: Solution by moving $y = -x + z$.

as this corner point is the one where the objective function passes through the feasible region. Therefore, we obtain the maximum value of the objective function $z = \frac{6}{5} + 3 = \frac{21}{5}$, as shown in Figure 2.6.

In this example, we have two decision variables, so the objective function is expressed by a line and the feasible region is expressed by a two-dimensional area surrounded by several lines. In the case of three decision variables, a three-dimensional space is considered. The objective function is expressed by a plane and the feasible region is expressed by a space surrounded by some planes associated with their constraints. For the general case of $n$ decision variables, an $n$-dimensional space is considered. The objective function is expressed by a hyperplane, and the feasible region is expressed by a space surrounded by some hyperplanes associated with their constraints.

Let us consider how to obtain an optimum solution of an LP problem in general. In an LP problem, if the problem has an optimum solution and at

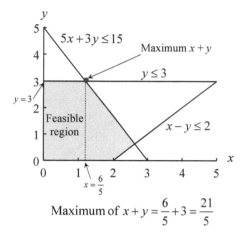

$$\text{Maximum of } x + y = \frac{6}{5} + 3 = \frac{21}{5}$$

Figure 2.6: Result by moving $y = -x + z$.

least one corner point of the feasible region exists, an optimum solution is one of the corner points. Therefore, we are able to get the optimum solution by checking each value of the objective function associated with every corner point.

Figure 2.7 shows every corner point. There are five corner points, $(0,0)$, $(0,3)$, $(\frac{6}{5},3)$, $(\frac{21}{8},\frac{5}{8})$, and $(2,0)$. Table 2.2 shows all the values of $x + y$ for every corner point. At the corner point of $(\frac{6}{5},3)$, $x + y = \frac{21}{5} = 4.2$ is the maximum value among the values of $x + y$ associated with every corner point.

## 2.3    Simplex method

If the number of decision variables and the number of constraints becomes large, the complexity of obtaining all the corner points and their corresponding values of the objective function is significant. This makes the computation times so long that the solution cannot be obtained in a practical time. To solve this issue, a more efficient way of finding the optimum solution for an LP problem, called the simplex method, was invented by Dantzig.

The simplex method uses the idea that at least one of the corner points is the optimum solution. We select one of the corner points as a starting point. We then walk along the boundary lines. In the case of a maximizing problem, we walk along the paths on which the value of the objection function does not decrease. In the case of a minimizing problem, we walk along paths on which the value of the objection function does not increase. In other words, if we cannot find a path that improves the value of the objective function at a certain corner point, the corner point is the optimum solution.

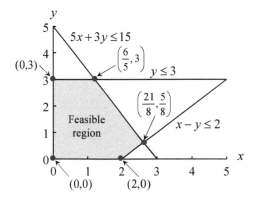

Figure 2.7: Corner points of feasible region.

Table 2.2: Values of $x + y$ for every corner point.

| Corner point $(x, y)$ | $x + y$ |
| --- | --- |
| $(0, 0)$ | $0$ |
| $(0, 3)$ | $3$ |
| $\left(\frac{6}{5}, 3\right)$ | $\frac{21}{5} = 4.2$ |
| $\left(\frac{21}{8}, \frac{5}{8}\right)$ | $\frac{13}{4} = 3.25$ |
| $(2, 0)$ | $2$ |

Figure 2.7 has five boundaries: $(0, 0) \leftrightarrow (0, 3)$, $(0, 3) \leftrightarrow \left(\frac{6}{5}, 3\right)$, $\left(\frac{6}{5}, 3\right) \leftrightarrow \left(\frac{21}{8}, \frac{5}{8}\right)$, $\left(\frac{21}{8}, \frac{5}{8}\right) \leftrightarrow (2, 0)$, and $(2, 0) \leftrightarrow (0, 0)$. Let us start at the corner point of $(0, 0)$. We have two possible paths, $(0, 0) \to (0, 3)$ and $(0, 0) \to (2, 0)$, from $(0, 0)$. At the corner point of $(0, 0)$, the value of the objective function is 0. At the corner points of $(0, 3)$ and $(2, 0)$, the values of the objective function are 3 and 2, respectively. The values of the objective function of both paths are increased. Therefore, we can choose either path. In Figure 2.8, we choose path $(0, 0) \to (0, 3)$, and $(0, 3) \to \left(\frac{6}{5}, 3\right)$ is considered. The value of the objective function at the corner point of $\left(\frac{6}{5}, 3\right)$ is 4.2, which is increased compared with the value of 3 associated with $(0, 3)$. If we continue to move from $\left(\frac{6}{5}, 3\right)$ to $\left(\frac{21}{8}, \frac{5}{8}\right)$, the value of the objective function, which is 3.25, decreases. Therefore, the corner point of $\left(\frac{6}{5}, 3\right)$ gives the maximum solution, where the value of the objective function is 4.2.

From the starting corner point of $(0, 0)$, Figure 2.9 shows the case that the other path, $(0, 0) \to (2, 0)$, is selected; in Figure 2.8, $(0, 0) \to (0, 3)$ is selected. The value of the objective function at $(0, 2)$ is 2. We keep moving to

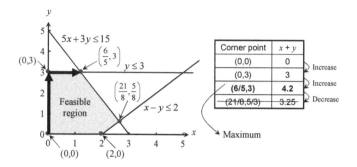

Figure 2.8: Example of simplex method. Search for boundary route of $(0,0) \leftrightarrow (0,3) \leftrightarrow \left(\frac{6}{5},3\right)$.

the corner points $\left(\frac{21}{8}, \frac{5}{8}\right)$ and $\left(\frac{6}{5}, 3\right)$. The values of the objective function are 3.25 and 4.2, respectively. They are increasing, so that we keep moving to the next corner point, $(0,3)$. However, the value of the objective function at this corner point is 3, which is decreased. Therefore, the maximum solution is the corner point of $\left(\frac{6}{5}, 3\right)$, where the value of the objective function is 4.2. The maximum solution in Figure 2.9 is the same as that in Figure 2.8.

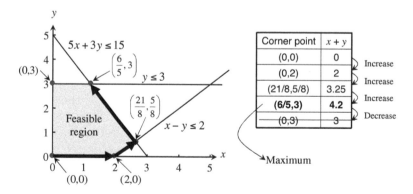

Figure 2.9: Example of simplex method. Search for boundary route of $(0,0) \to (2,0) \to \left(\frac{21}{8}, \frac{5}{8}\right) \to \left(\frac{6}{5}, 3\right)$.

Let us consider another example LP problem. A small factory uses a machine to produce two products: bread and noodles. How much powder is used to produce bread and noodles so as to maximize the profit per day? With 1 kilogram of powder, the profits of bread and noodles are \$5 and \$3, respectively. The machine is not able to produce more than one product at the same time. Both products are made with the same material, powder. The holding bin can keep only 80 kilograms of powder per day. The maximum runtime for the machine is 20 hours per day. The average times to make bread and noodles are 30 and 10 minutes per kilogram, respectively.

As the decision variables to this optimization problem, let $x$ be the weight of powder (kg) to produce bread, and $y$ be the weight of powder (kg) to produce noodle. This problem is formulated as an LP problem as follows:

$$\text{Objective} \quad \max \quad 5x + 3y \tag{2.18a}$$

$$\text{Constraints} \quad x + y \le 80 \tag{2.18b}$$

$$30x + 10y \le 1200 \tag{2.18c}$$

$$x \ge 0 \tag{2.18d}$$

$$y \ge 0. \tag{2.18e}$$

Eq. (2.18a) indicates the total profit of bread and noodle per 1 kg. We want to maximize this profit. Eq. (2.18b) indicates the constraint of the limitation of the powder. The powder used to produce either product must not exceed 80 kg. Eq. (2.18c) indicates the constraint of the total production time, which must be less than 20 hours, $20 \times 60 = 1200$ minutes. Eqs. (2.18d) and (2.18e) indicate non-negative values of $x$ and $y$. The constraints of Eqs. (2.18b)–(2.18e) are plotted in Figure 2.10.

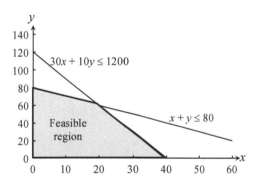

Figure 2.10: Feasible region.

We use the simplex method to solve this problem, as shown in Figure 2.11. Let us start at the corner point of $(0,0)$. The value of the objective function is 0. Here, we have two paths, $(0,0) \to (0,80)$ and $(0,0) \to (40,0)$. The values of the objective function are 240 and 200, respectively. They are increasing, so that we can select either of them to move. Let us move to the corner point of

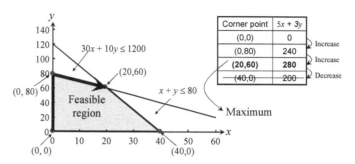

Figure 2.11: Example of simplex method. Search for boundary route of $(0, 80) \leftrightarrow (20, 60)$.

$(0, 80)$. The next corner point is $(20, 60)$. The value of the objective function at this corner point is 280, which is increased compared with the value of 240 associated with $(0, 80)$. We then move to the next corner point: $(40, 0)$. At this corner point, the value of the objective function is 200, a decrease. Therefore, the maximum solution is at the corner point of $(20, 60)$, where the value of the objective function is 280. We conclude that making 20 kg of bread and 60 kg of noodles achieves the maximum profit, \$280 per day.

## 2.4 Dual problem

Considering the problem of the bread and noodle factory, Eqs. (2.18a)–(2.18e), let us change the name of the decision variables from $x$ to $x_1$ and from $y$ to $x_2$ to deal with the problem in a more general manner. In other words, $x_1$ is the weight of powder (kg) to produce bread and $x_2$ is the weight of powder (kg) to produce noodles. We rewrite the problem into an LP problem formulation as follows:

$$\text{Objective} \quad \max \quad z = 5x_1 + 3x_2 \tag{2.19a}$$
$$\text{Constraints} \quad x_1 + x_2 \leq 80 \tag{2.19b}$$
$$30x_1 + 10x_2 \leq 1200 \tag{2.19c}$$
$$x_1 \geq 0 \tag{2.19d}$$
$$x_2 \geq 0. \tag{2.19e}$$

In agreement with the prior solution, the maximum solution is $(x_1, x_2) = (20, 60)$, and $z = 280$.

Here, we introduce new non-negative variables, $y_1 \ (\geq 0)$ and $y_2 \ (\geq 0)$. Let us multiply Eq. (2.19b) by $y_1$, and Eq. (2.19c) by $y_2$; the following equations are yielded:

$$x_1 y_1 + x_2 y_1 \leq 80 y_1 \tag{2.20}$$
$$30 x_1 y_2 + 10 x_2 y_2 \leq 1200 y_2. \tag{2.21}$$

After we sum Eqs. (2.20) and (2.21), we obtain

$$(y_1 + 30 y_2) x_1 + (y_1 + 10 y_2) x_2 \leq 80 y_1 + 1200 y_2. \tag{2.22}$$

By comparing Eq. (2.22) to the objective function, Eq. (2.19a), we can assume

$$y_1 + 30 y_2 \geq 5 \tag{2.23}$$
$$y_1 + 10 y_2 \geq 3. \tag{2.24}$$

From Eqs. (2.22), (2.23), and (2.24), we obtain

$$5 x_1 + 3 x_2 \leq (y_1 + 30 y_2) x_1 + (y_1 + 10 y_2) x_2 \leq 80 y_1 + 1200 y_2. \tag{2.25}$$

From Eqs. (2.22) and (2.23), the upper bound of $5 x_1 + 3 x_2$ is $80 y_1 + 1200 y_2$. To minimize the upper bound $80 y_1 + 1200 y_2$, we consider the following LP problem.

| | | | |
|---|---|---|---|
| Objective | min | $w = 80 y_1 + 1200 y_2$ | (2.26a) |
| Constraints | | $y_1 + 30 y_2 \geq 5$ | (2.26b) |
| | | $y_1 + 10 y_2 \geq 3$ | (2.26c) |
| | | $y_1 \geq 0$ | (2.26d) |
| | | $y_2 \geq 0$ | (2.26e) |

by solving this problem, the optimum solution is obtained as $(y_1, y_2) = (2, 0.1)$, and $w = 280$. The solution is the same as that yielded by Eqs. (2.18a)–(2.18e).

The problem of Eqs. (2.18a)–(2.18e) is the dual problem of Eqs. (2.26a)–(2.26e), and vice versa. The main problem is called the primal problem and the other is called the dual problem; $y_1$ and $y_2$ are called dual variables. In general, if one LP problem is the dual problem of another LP problem, both objective functions have the same optimum value, as will be described later.

We can explain the problem of Eqs. (2.26a)–(2.26e) as follows: $y_1 (\geq 0)$ is the cost per kilogram of powder to produce bread or noodles, \$/kg; $y_2 (\geq 0)$ is cost to run the machine per minute, \$/min. The objective function, $w = 80 y_1 + 1200 y_2$, is the total cost for using the powder and running the machine. We want to minimize the total cost. Eq. (2.26b) indicates that the cost of using 1 kg of powder and running the machine to produce bread is $y_1 [\$/kg] + 30 [\min/kg] \times y_2 [\$/\min]$, and this cost must not be lower than \$5/kg, which is the profit of bread when 1 kg of powder is used. Eq. (2.26c) indicates that the cost of using 1 kg of powder and running the machine to produce noodles is $y_1 [\$/kg] + 10 [\min/kg] \times y_2 [\$/\min]$, and this must not be lower than \$3/kg, which is the profit of noodles when 1 kg of powder is used.

The relationship between the primal problem and the dual problem is explained as follows. Eqs. (2.27a)–(2.27c) show the primal LP problem as

$$\text{Objective} \quad \max \quad \boldsymbol{c}^T \boldsymbol{x} \tag{2.27a}$$

$$\text{Constraints} \quad \boldsymbol{A}\boldsymbol{x} \leq \boldsymbol{b} \tag{2.27b}$$

$$\boldsymbol{x} \geq 0, \tag{2.27c}$$

where

$$\boldsymbol{x}^T = [x_1, \ldots, x_n] \tag{2.28a}$$

$$\boldsymbol{b}^T = [b_1, \ldots, b_m] \tag{2.28b}$$

$$\boldsymbol{c}^T = [c_1, \ldots, c_n] \tag{2.28c}$$

$$\boldsymbol{A} = \begin{bmatrix} a_{11} & a_{12} & \cdots & a_{1n} \\ a_{21} & a_{22} & \cdots & a_{2n} \\ \vdots & \vdots & \ddots & \vdots \\ a_{m1} & a_{m2} & \cdots & a_{mn} \end{bmatrix}, \tag{2.28d}$$

where $\boldsymbol{x}$ are the decision variables, and $\boldsymbol{A}$, $\boldsymbol{b}$, and $\boldsymbol{c}$ are parameters. The dual problem of Eqs. (2.27a)–(2.27c) is represented by

$$\text{Objective} \quad \min \quad \boldsymbol{b}^T \boldsymbol{y} \tag{2.29a}$$

$$\text{Constraints} \quad \boldsymbol{A}^T \boldsymbol{y} \geq \boldsymbol{c} \tag{2.29b}$$

$$\boldsymbol{y} \geq 0, \tag{2.29c}$$

where

$$\boldsymbol{y}^T = [y_1, \ldots, y_m]. \tag{2.30}$$

In the dual problem, $\boldsymbol{y}$ are decision variables, and $\boldsymbol{A}$, $\boldsymbol{b}$, and $\boldsymbol{c}$ are the same parameters as the primal problem.

In the dual theorem, for a pair of primal and dual problems, if there is an optimum solution of either the primal problem or the dual problem, it is guaranteed that an optimum solution of the other problem exists. Moreover, both optimum values of the objective functions are the same.

## 2.5   Integer linear programming problem

An LP problem in which decision variables take only integer values is called an integer linear programming (ILP) problem. In the previous problems, decision variables were considered real numbers and non-negative values. Some problems need only integer values as decision variables, such as the number of people or the number of pieces.

The methods described in Sections 2.2 and 2.3 are not able to be applied as they were designed to solve ILP problems. An LP problem, in which the decision variables include both integer values and real values, is called a mixed integer linear programming (MILP) problem.

In general, it takes more time to solve an ILP problem than an LP one. Let us consider Eqs. (2.17a)–(2.17f) again, and assume that the decision variables are limited to integer values:

$$\text{Objective} \quad \max \quad x + y \tag{2.31a}$$

$$\text{Constraints} \quad 5x + 3y \leq 15 \tag{2.31b}$$

$$x - y \leq 2 \tag{2.31c}$$

$$y \leq 3 \tag{2.31d}$$

$$x = 0, 1, \cdots \text{(integer value)} \tag{2.31e}$$

$$y = 0, 1, \cdots \text{(integer value)}. \tag{2.31f}$$

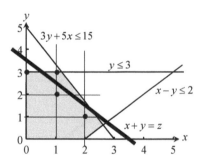

Figure 2.12: Feasible region of integer linear programming problem.

In an LP problem, at least one of the corner points is the optimum solution. Therefore, we need check only the corner points to determine the optimum solution. However, in an ILP problem, we have to check every possible grid point in the feasible region to identify the optimum value of the objective function, as shown in Figure 2.12. In Figure 2.12, we need to check the value of the objective function of four grid points: $(0, 3)$, $(1, 2)$, $(1, 3)$, and $(2, 1)$. We find that the optimum solution is $(1, 3)$, and that the maximum value of the objective function is 4.

Let us consider a large-scale ILP problem as follows:

$$\begin{array}{llr}
\text{Objective} & \max \quad x + y & \text{(2.32a)} \\
\text{Constraints} & 5x + 3y \leq 1500 & \text{(2.32b)} \\
& x - y \leq 200 & \text{(2.32c)} \\
& y \leq 300 & \text{(2.32d)} \\
& x = 0, 1, \cdots \text{(integer value)} & \text{(2.32e)} \\
& y = 0, 1, \cdots \text{(integer value)}. & \text{(2.32f)}
\end{array}$$

Figure 2.13 shows that we need to find the values of the objective function by considering several grid points. The optimum solution is the grid point of $(120, 300)$, and the maximum value of objective function is 420. This problem takes some time to calculate the solution because we have to consider many more grid points than those in Figure 2.12.

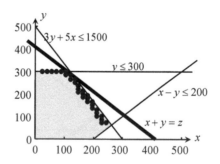

Figure 2.13: Image for feasible region of large-scale integer linear programming problem.

## Exercise 2.1

Solve the following LP problem.

$$\begin{array}{ll}
\text{Objective} & \max \quad 8x_1 + 6x_2 \\
\text{Constraints} & 2x_1 + x_2 \leq 30 \\
& x_1 + 2x_2 \leq 24 \\
& x_1 \geq 0 \\
& x_2 \geq 0
\end{array}$$

## Exercise 2.2

Solve the following LP problem.

$$
\begin{aligned}
\text{Objective} \quad & \max \quad 10x_1 + 12x_2 \\
\text{Constraints} \quad & 2x_1 + 3x_2 \leq 30 \\
& 3x_1 + 2x_2 \leq 24 \\
& x_1 \geq 0 \\
& x_2 \geq 0
\end{aligned}
$$

## Exercise 2.3

Solve the following LP problem.

$$
\begin{aligned}
\text{Objective} \quad & \min \quad 80x_1 + 1200x_2 \\
\text{Constraints} \quad & x_1 + 30x_2 \geq 5 \\
& x_1 + 10x_2 \geq 3 \\
& x_1 \geq 0 \\
& x_2 \geq 0
\end{aligned}
$$

## Exercise 2.4

Solve the following ILP problem.

$$
\begin{aligned}
\text{Objective} \quad & \max \quad 10x_1 + 12x_2 \\
\text{Constraints} \quad & 2x_1 + 3x_2 \leq 30 \\
& 3x_1 + 2x_2 \leq 24 \\
& x_1 = 0, 1, \cdots \text{integer value} \\
& x_2 = 0, 1, \cdots \text{integer value}
\end{aligned}
$$

## Exercise 2.5

Solve the following ILP problem.

$$
\begin{aligned}
\text{Objective} \quad & \max \quad 12x_1 + 10x_2 \\
\text{Constraints} \quad & 2x_1 + 3x_2 \leq 30 \\
& 3x_1 + 2x_2 \leq 24 \\
& x_1 = 0, 1, \cdots \text{integer value} \\
& x_2 = 0, 1, \cdots \text{integer value}
\end{aligned}
$$

# Chapter 3

# GLPK (GNU Linear Programming Kit)

This chapter introduces the software sufficient to solve LP problems. This software is called the GNU linear programming kit (GLPK). It is easy to analytically solve an LP problem with a few decision variables by hand. However, if the number of decision variables increases, the manual approach becomes infeasible. Therefore, a practical tool is required to solve LP problems using computers. Both commercial and free programs have been released to solve LP problems. In this book, the free software, named GLPK, is used as an LP solver.

## 3.1 How to obtain GLPK and install it

GLPK is freely available. GLPK is an open-source software package to solve LP problems, developed by Andrew O. Makhorin. It is a set of routines written in ANSI C and organized in the form of a callable library. GLPK supports the GNU MathProg modeling language, which is a subset of AMPL (a modeling language for mathematical programming). The language is supported by popular commercial mathematical programming solvers, for example, CPLEX®. Once readers understand how to solve mathematical programming problems on communication networks using GLPK in this book, they will also be able to easily tackle similar problems by applying other solvers. GLPK is available for download from this website:

http://www.gnu.org/s/glpk/

This book uses GLPK version 4.45. The installation manual can be found on the website. After installation and configuring GLPK path, use the command 'glpsol --version' to show GLPK version. The information of GLPK is shown as Listing 3.1.

Listing 3.1: Console

```
1  $ glpsol --version
2  GLPSOL: GLPK LP/MIP Solver, v4.45
3
4  Copyright (C) 2000, 2001, 2002, 2003, 2004, 2005, 2006, 2007, 2008,
5  2009, 2010 Andrew Makhorin, Department for Applied Informatics, Moscow
6  Aviation Institute, Moscow, Russia. All rights reserved.
7
8  This program has ABSOLUTELY NO WARRANTY.
9
10 This program is free software; you may re-distribute it under the terms
11 of the GNU General Public License version 3 or later.
```

## 3.2   Usage of GLPK

Let us solve an LP problem involving Eqs. (2.17a)–(2.17f) by GLPK. A file that describes the optimization problem is called a model file. Listing 3.2 shows the model file of the LP problem for Eqs. (2.17a)–(2.17f).

Listing 3.2: Model file: lp-ex1.mod

```
1  /* lp-ex1.mod */
2
3  var x >= 0 ;
4  var y >= 0 ;
5
6  maximize z: x + y ;
7  s.t. st1:  5*x + 3*y <= 15 ;
8  s.t. st2:  x - y <= 2 ;
9  s.t. st3:  y <=3 ;
10
11 end ;
```

To solve this problem with GLPK, we type the command line 'glpsol', as shown at line 1 in Listing 3.3. In this command line, '-m' is an option indicating that 'lp-ex1.mod' is the model file (Listing 3.2). Option '-o' indicates the output file lp-ex1.out (Listing 3.4). After running this command, GLPK will report as shown in Listing 3.3.

Listing 3.3: Console

```
1  $ glpsol -m lp-ex1.mod -o lp-ex1.out
2  GLPSOL: GLPK LP/MIP Solver, v4.45
3  Parameter(s) specified in the command line:
4   -m lp-ex1.mod -o lp-ex1.out
5  Reading model section from lp-ex1.mod...
6  11 lines were read
7  Generating z...
8  Generating st1...
9  Generating st2...
10 Generating st3...
11 Model has been successfully generated
12 GLPK Simplex Optimizer, v4.45
13 4 rows, 2 columns, 7 non-zeros
14 Preprocessing...
15 2 rows, 2 columns, 4 non-zeros
16 Scaling...
17  A: min|aij| =  1.000e+00  max|aij| =  5.000e+00  ratio =  5.000e+00
18 Problem data seem to be well scaled
```

```
Constructing initial basis...
Size of triangular part = 2
*      0: obj =    0.000000000e+00  infeas =   0.000e+00 (0)
*      3: obj =    4.200000000e+00  infeas =   0.000e+00 (0)
OPTIMAL SOLUTION FOUND
Time used:    0.0 secs
Memory used: 0.1 Mb (108176 bytes)
Writing basic solution to 'lp-ex1.out'...
```

The above report shows the generation of the objective function and constraints, the optimization process by GLPK, and the determination of the optimum solution. The optimum solution and the value of the objective function are shown in output file '`lp-ex1.out`' (Listing 3.4).

Listing 3.4: Output file: lp-ex1.out

```
Problem:    lp
Rows:       4
Columns:    2
Non-zeros:  7
Status:     OPTIMAL
Objective:  z = 4.2 (MAXimum)

   No.   Row name   St   Activity     Lower bound   Upper bound    Marginal
------ ------------ -- ------------- ------------- ------------- ----------
     1 z            B           4.2
     2 st1          NU           15                          15         0.2
     3 st2          B          -1.8                           2
     4 st3          NU            3                           3         0.4

   No. Column name  St   Activity     Lower bound   Upper bound    Marginal
------ ------------ -- ------------- ------------- ------------- ----------
     1 x            B           1.2             0
     2 y            B             3             0

Karush-Kuhn-Tucker optimality conditions:

KKT.PE: max.abs.err = 2.22e-16 on row 1
        max.rel.err = 2.36e-17 on row 1
        High quality

KKT.PB: max.abs.err = 0.00e+00 on row 0
        max.rel.err = 0.00e+00 on row 0
        High quality

KKT.DE: max.abs.err = 5.55e-17 on column 1
        max.rel.err = 1.85e-17 on column 1
        High quality

KKT.DB: max.abs.err = 0.00e+00 on row 0
        max.rel.err = 0.00e+00 on row 0
        High quality

End of output
```

Lines 1–5 in Listing 3.4 show the information of the optimum problem. Line 6 shows that the maximum value of the objective function is 4.2. Lines 8–13 show the information of the objective function and constraints. Lines 15–18 show the information of the optimum values of the decision variables. Values in the column 'Activity' in lines 17 and 18 show that the optimum value of $(x, y)$ is $(1.2, 3)$.

Listing 3.5 shows the model file for the LP problem of Eqs. (2.18a)–(2.18e).

Listing 3.5: Model file: lp-ex2.mod

```
1   /* lp-ex2.mod */
2
3   var y1 >= 0 ;
4   var y2 >= 0 ;
5
6   minimize w: 80*y1 + 1200*y2 ;
7   s.t. st1: y1 + 30*y2 >= 5 ;
8   s.t. st2: y1 + 10*y2 >= 3 ;
9
10  end ;
```

We run GLPK by the command line `glpsol -m lp-ex2.mod -o lp-ex2.out`. The optimum solution is $(y_1, y_2) = (2, 0.1)$, and the value of the objective function is $w = 280$.

## Exercise 3.1

A factory produces health food. There are four raw materials, A, B, C, and D, that can be used to produce the food. At least 18 kg of protein, 31 kg of carbohydrate, and 25 kg of fat are required to produce the food. Ingredients of each raw material are shown in Table 3.1.

Table 3.1: Ingredients of each raw material.

| Raw materials | Nutrient ratio | | | Price ($/kg) |
|---|---|---|---|---|
| | Protein | Carbohydrate | Fat | |
| A | 0.18 | 0.43 | 0.31 | 5.00 |
| B | 0.31 | 0.25 | 0.37 | 7.50 |
| C | 0.12 | 0.12 | 0.37 | 3.75 |
| D | 0.18 | 0.50 | 0.12 | 2.50 |

1. The factory wants to produce the food while minimizing the nutrient cost. Formulate an LP problem.

2. Solve the LP problem.

3. Let this above problem be a primal problem. Formulate the dual problem.

4. Solve the dual problem and compare the solution with that of the primal problem.

## Exercise 3.2

A factory produces three hair treatment products, regular shampoo, exclusive shampoo, and conditioner. They use the basic chemicals, A, B, and C, in specified proportions, as shown in Table 3.2. The profit values of regular shampoo, exclusive shampoo, and conditioner are 1.5, 2.0, and 2.5 \$/liter, respectively. The factory wants to sell at least 30 liters of exclusive shampoo.

Table 3.2: Raw materials for hair treatment products.

| Raw material | Ratio of basic chemicals used for hair treatment products | | | Quantity in stock (liters) |
| --- | --- | --- | --- | --- |
| | Regular shampoo | Exclusive shampoo | Conditioner | |
| A | 0.3 | 0.5 | 0.2 | 100 |
| B | 0.6 | 0.3 | 0.1 | 150 |
| C | 0.1 | 0.2 | 0.7 | 200 |

1. The factory wants to get the maximum profit. Formulate the problem as an LP problem.

2. Solve the LP problem.

3. Let the above problem be a primal problem. Formulate the dual problem.

4. Solve the dual problem and compare the solution with that of the primal problem.

# Chapter 4

# Basic problems for communication networks

This chapter describes basic problems posed by communication networks that can be tackled by linear programming. Formulations, solutions by GLPK, and related algorithms for various problems are presented in this chapter.

## 4.1 Shortest path problem

### 4.1.1 Linear programming problem

The network is represented by directed graph $G(V, E)$, where $V$ is the set of vertices (nodes) and $E$ is the set of links. A link from node $i$ to node $j$ is expressed by $(i, j) \in E$; $d_{ij}$ is the link cost of $(i, j)$; $x_{ij}^{pq}$, where $0 \leq x_{ij}^{pq} \leq 1$, is the traffic volume from node $p \in V$ to node $q \in V$ routed through $(i, j) \in E$. In Figure 4.1, considering node 1 as a source node ($p = 1$) and node 4 as a destination node ($q = 4$), we want to find the shortest path from node 1 to node 4. The shortest path problem is formulated as the following LP problem.

$$\text{Objective} \quad \min \quad 5x_{12} + 8x_{13} + 2x_{23} + 7x_{24} + 4x_{34} \quad \text{(4.1a)}$$

$$\text{Constraints} \quad x_{12} + x_{13} = 1 \quad \text{(4.1b)}$$

$$x_{12} - x_{23} - x_{24} = 0 \quad \text{(4.1c)}$$

$$x_{13} + x_{23} - x_{34} = 0 \quad \text{(4.1d)}$$

$$0 \leq x_{12} \leq 1 \quad \text{(4.1e)}$$

$$0 \leq x_{13} \leq 1 \quad \text{(4.1f)}$$

$$0 \leq x_{23} \leq 1 \quad \text{(4.1g)}$$

$$0 \leq x_{24} \leq 1 \quad \text{(4.1h)}$$

$$0 \leq x_{34} \leq 1 \quad \text{(4.1i)}$$

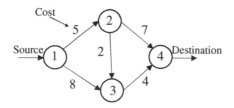

Figure 4.1: Network model of shortest path problem. Number on each link represents link cost.

The decision variables are $x_{12}$, $x_{13}$, $x_{23}$, $x_{24}$, and $x_{34}$. Eq. (4.1a) expresses the objective function to find the minimum path cost from node 1 to node 4. Eq. (4.1a) indicates the objective function, which is the sum of the costs of links along the path(s). In Eq. (4.1a), the costs of links that are not part of the path(s) are not considered as the value of $x_{ij}$ is 0 when $(i, j)$ is not on the path(s). Eqs. (4.1c)–(4.1i) shows the constraints. Eqs. (4.1b)–(4.1d) indicate flow conservation. Eq. (4.1b) maintains the condition of flows at the source node, node 1. The outgoing traffic volume from node 1, $x_{12} + x_{13}$, is equal to 1, which is the incoming traffic volume, as shown in Figure 4.2(a). Node 1 transmits the traffic volume of 1. Eq. (4.1c) maintains the condition of flows at an intermediate node, node 2. It expresses the fact that the incoming traffic volume of node 2, $x_{12}$, is equal to the outgoing traffic volume of node 2, $x_{23} + x_{24}$, as shown in Figure 4.2(b). Eq. (4.1d) maintains the condition of flows at an intermediate node, node 3. It expresses the fact that the incoming traffic volume of node 3, $x_{13} + x_{23}$, is equal to the outgoing traffic volume of node 3, $x_{34}$, as shown in Figure 4.2(c). Eqs. (4.1e)–(4.1i) present the ranges of $x_{ij}$.

At the destination node, node 4, the condition that maintains flows, is

$$x_{24} + x_{34} = 1, \tag{4.2}$$

as shown in Figure 4.2(d). However, Eq. (4.2) is obtained by using Eqs. (4.1b)–(4.1d). Therefore, Eq. (4.2) is always guaranteed if Eqs. (4.1b)–(4.1d) are satisfied. Therefore, Eqs. (4.1b)–(4.1d) are enough to maintain flow conservation.

Let us solve the LP problem presented in Eqs. (4.1a)–(4.1i) using GLPK. The model file for this problem is shown in Listing 4.1.

Listing 4.1: Model file: sp-ex1.mod

```
1  /* sp-ex1.mod */
2
3  /* Decision variables */
4  var x12 <=1, >=0 ;
5  var x13 <=1, >=0 ;
6  var x23 <=1, >=0 ;
7  var x24 <=1, >=0 ;
8  var x34 <=1, >=0 ;
9
10 /* Objective function */
```

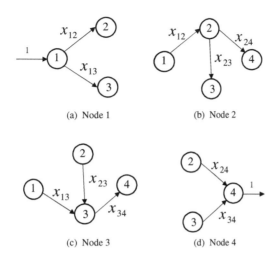

(a) Node 1                          (b) Node 2

(c) Node 3                          (d) Node 4

Figure 4.2: Input/output of traffic at each node.

```
minimize PATH_COST: 5*x12 + 8*x13 + 2*x23 + 7*x24 + 4*x34 ;

/* Constraints */
s.t. NODE1: x12 + x13 = 1 ;
s.t. NODE2: x12 - x23 - x24 = 0 ;
s.t. NODE3: x13 + x23 - x34 = 0 ;

end;
```

Lines 4–8 define the decision variables, $x_{ij}$, for $(i, j)$, and express the ranges for Eqs. (4.1e)–(4.1i). Line 11 expresses Eq. (4.1a), where the objective function is minimized. Lines 14–16 express the constraints as presented in Eqs. (4.1b)–(4.1d).

To solve this problem with GLPK, we run command line 'glpsol' as shown on line 1 in Listing 4.2. In this command line, '-m' is an option to indicate that 'sp-ex1.mod' is the model file (Listing 4.1). Option '-o' indicates the output file **sp-ex1.out** (Listing 4.3). If we run this command, GLPK reports Listing 4.2.

Listing 4.2: Console

```
$ glpsol -m sp-ex1.mod -o sp-ex1.out
GLPSOL: GLPK LP/MIP Solver, v4.45
Parameter(s) specified in the command line:
 -m sp-ex1.mod -o sp-ex1.out
Reading model section from sp-ex1.mod...
19 lines were read
Generating PATH_COST...
Generating NODE1...
Generating NODE2...
Generating NODE3...
Model has been successfully generated
```

```
12  GLPK Simplex Optimizer, v4.45
13  4 rows, 5 columns, 13 non-zeros
14  Preprocessing...
15  3 rows, 3 columns, 6 non-zeros
16  Scaling...
17   A: min|aij| =  1.000e+00  max|aij| =  1.000e+00  ratio =  1.000e+00
18  Problem data seem to be well scaled
19  Constructing initial basis...
20  Size of triangular part = 3
21  *     0: obj =   1.200000000e+01  infeas =  0.000e+00 (0)
22  *     2: obj =   1.100000000e+01  infeas =  0.000e+00 (0)
23  OPTIMAL SOLUTION FOUND
24  Time used:   0.0 secs
25  Memory used: 0.1 Mb (108469 bytes)
26  Writing basic solution to 'sp-ex1.out'...
```

The above report describes the generation of the objective function and constraints, the optimization process by GLPK, and the identification of the optimum solution. The optimum solution and the value of the objective function are shown in the output file 'sp-ex1.out' (Listing 4.3).

Listing 4.3: Output file: sp-ex1.out

```
1   Problem:    sp
2   Rows:       4
3   Columns:    5
4   Non-zeros:  13
5   Status:     OPTIMAL
6   Objective:  PATH_COST = 11 (MINimum)
7
8      No.   Row name    St   Activity     Lower bound   Upper bound    Marginal
9    ------ ------------ -- ------------- ------------- ------------- -----------
10       1 PATH_COST    B             11
11       2 NODE1        NS             1             1                        =
12
13       3 NODE2        NS             0            -0                        =
12   -6
14       4 NODE3        NS             0            -0                        =
13   -4
14
15      No. Column name St   Activity     Lower bound   Upper bound    Marginal
16   ------ ------------ -- ------------- ------------- ------------- -----------
17       1 x12          NU             1             0             1
    -1
18       2 x13          B              0             0             1
19       3 x23          B              1             0             1
20       4 x24          NL             0             0             1
    1
21       5 x34          B              1             0             1
22
23   Karush-Kuhn-Tucker optimality conditions:
24
25   KKT.PE:  max.abs.err = 0.00e+00 on row 0
26            max.rel.err = 0.00e+00 on row 0
27            High quality
28
29   KKT.PB:  max.abs.err = 0.00e+00 on row 0
30            max.rel.err = 0.00e+00 on row 0
31            High quality
32
33   KKT.DE:  max.abs.err = 0.00e+00 on column 0
34            max.rel.err = 0.00e+00 on column 0
35            High quality
36
37   KKT.DB:  max.abs.err = 0.00e+00 on row 0
```

```
        max.rel.err = 0.00e+00 on row 0
        High quality

End of output
```

Lines 1–5 show the information of the optimum problem. Line 6 shows that the minimum value of the objective function is 11. Lines 8–13 show the information of the objective function and constraints. Lines 15–21 show the information of the optimum value of the decision variables. Values in the column 'Activity' in lines 17–21 show that the optimum solution is $x_{12} = 1$, $x_{13} = 0$, $x_{23} = 1$, $x_{24} = 0$, $x_{34} = 1$. In other words, the shortest path is $1 \to 2 \to 3 \to 4$, and the cost of this path is 11.

In general, the problem to find the shortest path is formulated as an LP problem:

$$\text{Objective} \quad \min \sum_{(i,j)\in E} d_{ij}x_{ij} \tag{4.3a}$$

$$\text{Constraints} \quad \sum_{j:(i,j)\in E} x_{ij} - \sum_{j:(j,i)\in E} x_{ji} = 1, \quad \text{if } i = p \tag{4.3b}$$

$$\sum_{j:(i,j)\in E} x_{ij} - \sum_{j:(j,i)\in E} x_{ji} = 0, \quad \forall i \neq p, q \in V \tag{4.3c}$$

$$0 \leq x_{ij} \leq 1, \quad \forall (i,j) \in E. \tag{4.3d}$$

$x_{ij}$ and $d_{ij}$ are the decision variable and the link cost of $(i,j)$, respectively. Eq. (4.3a) is the objective function that minimizes the path cost from node $p$ to node $q$. $x_{ij}$ is the traffic volume from node $p$ to node $q$ routed through $(i,j)$. Eqs. (4.3b)–(4.3d) are the constraints. Eqs. (4.3b)–(4.3c) express the conditions of the flow conservation. Eq. (4.3b) maintains the flows at the source node, node $p$. The difference between the incoming traffic volume and the outgoing traffic volume, $\sum_{j:(i,j)\in E} x_{ij} - \sum_{j:(j,i)\in E} x_{ji}$, is 1. Here, the outgoing traffic volume at node $p$ is 1. Eq. (4.3c) maintains flows at intermediate node $i$, where $i \neq p, q$. The outgoing traffic volume at node $i$, $\sum_{j:(i,j)\in E} x_{ij}$, is equal to the incoming traffic volume at node $i$, $\sum_{j:(j,i)\in E} x_{ji}$. Eq. (4.3d) is the range of $x_{ij}$.

At the destination node, node $q$, the condition to maintain flows is

$$\sum_{j:(i,j)\in E} x_{ij} - \sum_{j:(j,i)\in E} x_{ji} = -1, \text{if } i = q \tag{4.4}$$

Eq. (4.4) must be satisfied. However, Eq. (4.4) is deducted using Eqs. (4.3b)–(4.3c). Therefore, Eq. (4.4) is guaranteed by Eqs. (4.3b) and (4.3c), which is proved below.

**Proof:**

Eq. (4.3b) is written by

$$\sum_{j:(p,j)\in E} x_{pj} - \sum_{j:(j,p)\in E} x_{jp} = 1. \tag{4.5}$$

Eq. (4.3c) expresses a set of $N-2$ equations for $i \in V$, $i \neq p, q$. Let us take a sum over the left sides of Eq. (4.5) and $N-2$ equations expressed in Eq. (4.3c) and a sum over the right sides of them. As both sums are equal,

$$\sum_{j:(p,j)\in E} x_{pj} + \sum_{i\in V, i\neq p,q} \sum_{j:(i,j)\in E} x_{ij} - \sum_{j:(j,p)\in E} x_{jp} - \sum_{i\in V, i\neq p,q} \sum_{j:(j,i)\in E} x_{ji} = 1 \tag{4.6}$$

is obtained. Using the following relationships given by

$$\sum_{j:(p,j)\in E} x_{pj} + \sum_{i\in V, i\neq p,q} \sum_{j:(i,j)\in E} x_{ij} = \sum_{i\in V} \sum_{j:(i,j)\in E} x_{ij} - \sum_{j:(q,j)\in E} x_{qj}$$

and

$$\sum_{j:(j,p)\in E} x_{jp} + \sum_{i\in V, i\neq p,q} \sum_{j:(j,i)\in E} x_{ji} = \sum_{i\in V} \sum_{j:(j,i)\in E} x_{ji} - \sum_{j:(j,q)\in E} x_{jq},$$

Eq. (4.6) is transformed to

$$\sum_{i\in V} \sum_{j:(i,j)\in E} x_{ij} - \sum_{j:(q,j)\in E} x_{qj} - \sum_{i\in V} \sum_{j:(j,i)\in E} x_{ji} + \sum_{j:(j,q)\in E} x_{jq} = 1. \tag{4.7}$$

Using

$$\sum_{i\in V} \sum_{j:(i,j)\in E} x_{ij} - \sum_{i\in V} \sum_{j:(j,i)\in E} x_{ji} = 0,$$

Eq. (4.7) becomes

$$\sum_{j:(q,j)\in E} x_{qj} - \sum_{j:(j,q)\in E} x_{jq} = -1. \tag{4.8}$$

Eq. (4.8) is a simplification of Eq. (4.4).  ∎

As the model file describes `sp-ex1.mod` (Listing 4.1), it must be modified when some network conditions, including the topology and the link costs, are changed. It takes time to manually change the model file. Moreover, it is easy to make mistakes when modifying the model file. To avoid this issue, GLPK allows us to use a model file in conjunction with a separate input file. The general model is written in the model file, while the parameters or data including topology and link costs are written in the input file. If the network conditions are changed, we only have to modify the input file, without touching the model file.

Eqs. (4.3a)–(4.3d) are separately written in the model file as shown in Listing 4.4, and the input file as shown in Listing 4.5.

Listing 4.4: Model file: sp-gen.mod

```
/* sp-gen.mod */

/* Given parameters */
param N integer, >0;
param p integer, >0;
param q integer, >0;

set V := 1..N;
set E within {V,V};

param cost{E};

/* Decision variables */
var x{E} <= 1, >= 0;

/* Objective function */
minimize PATH_COST: sum{i in V} (sum{j in V} (cost[i,j]*x[i,j] ) ) ;

/* Constraints */
s.t. SOURCE{i in V: i = p && p != q}:
        sum{j in V} (x[i,j]) - sum{j in V}(x[j,i]) = 1;
s.t. INTERNAL{i in V: i != p && i != q && p != q }:
        sum{j in V} (x[i,j]) - sum{j in V}(x[j,i]) = 0;
end;
```

Lines 4, 5, 6, and 11 of the model file in Listing 4.4 define the types of parameters for the number of nodes, $N$, source node, $p$, destination node, $q$, and link costs.

Listing 4.5: Input file: sp-gen1.dat

```
/* sp-gen1.dat */

param p := 1;
param q := 4;
param N := 4;

param : E : cost :=
1 1 100000
1 2 5
1 3 8
1 4 100000
2 1 100000
2 2 100000
2 3 2
2 4 7
3 1 100000
3 2 100000
3 3 100000
3 4 4
4 1 100000
4 2 100000
4 3 100000
4 4 100000
;
end;
```

Lines 3–5 of the input file in Listing 4.5 define the values of parameters $p$, $q$, and $N$. Lines 8–23 define the link cost of $(i, j)$. To handle the case of two nodes with no link between them (e.g., $(1, 1)$ and $(1, 4)$), the cost for $(i, j)$ is set to a large enough number that the pair will never be considered in forming the shortest path. In this case, we set the cost to 10000.

We run a command line 'glpsol' as shown on line 1 in Listing 4.6. The model file, the input file, and the output file are specified by options. Option '-m' indicates model file sp-gen.mod, as shown in Listing 4.4. Option '-d' indicates the corresponding input file sp-gen1.dat, as shown in Listing 4.5. Option '-o' indicates the output file sp-gen1.out. After running the command, GLPK will report Listing 4.6. The result is the same as the program in Listing 4.1. The output file sp-gen1.out is shown in Listing 4.7.

Listing 4.6: Console: sp-gen1.txt

```
1  $ glpsol -m sp-gen.mod -d sp-gen1.dat -o sp-gen1.out
2  GLPSOL: GLPK LP/MIP Solver, v4.45
3  Parameter(s) specified in the command line:
4   -m sp-gen.mod -d sp-gen1.dat -o sp-gen1.out
5  Reading model section from sp-gen.mod...
6  25 lines were read
7  Reading data section from sp-gen1.dat...
8  25 lines were read
9  Generating PATH_COST...
10 Generating SOURCE...
11 Generating INTERNAL...
12 Model has been successfully generated
13 GLPK Simplex Optimizer, v4.45
14 4 rows, 16 columns, 34 non-zeros
15 Preprocessing...
16 3 rows, 9 columns, 15 non-zeros
17 Scaling...
18  A: min|aij| =  1.000e+00  max|aij| =  1.000e+00  ratio =  1.000e+00
19 Problem data seem to be well scaled
20 Constructing initial basis...
21 Size of triangular part = 3
22 *      0: obj =   1.000000000e+05  infeas =  0.000e+00 (0)
23 *      6: obj =   1.100000000e+01  infeas =  0.000e+00 (0)
24 OPTIMAL SOLUTION FOUND
25 Time used:   0.0 secs
26 Memory used: 0.1 Mb (125211 bytes)
27 Writing basic solution to 'sp-gen1.out'...
```

Listing 4.7: Output file: sp-gen1.out

```
1  Problem:    sp
2  Rows:       4
3  Columns:    16
4  Non-zeros:  34
5  Status:     OPTIMAL
6  Objective:  PATH_COST = 11 (MINimum)
7
8     No.   Row name    St   Activity     Lower bound   Upper bound    Marginal
9   ------ ------------ -- ------------- ------------- ------------- -------------
10     1 PATH_COST    B             11
11     2 SOURCE[1]    NS             1             1                       =
12
12     3 INTERNAL[2]  NS             0            -0                       =
7
13     4 INTERNAL[3]  NS             0            -0                       =
4
14
15     No. Column name  St   Activity     Lower bound   Upper bound    Marginal
16   ------ ------------ -- ------------- ------------- ------------- -------------
17     1 x[1,1]       NL             0             0             1        100000
18     2 x[1,2]       B              1             0             1
19     3 x[1,3]       B              0             0             1
20     4 x[1,4]       NL             0             0             1         99988
```

```
    5  x[2,1]      NL          0           0          1      100005|
    6  x[2,2]      NL          0           0          1      100000|
    7  x[2,3]      NU          1           0          1          -1|
    8  x[2,4]      B           0           0          1
    9  x[3,1]      NL          0           0          1      100008|
   10  x[3,2]      NL          0           0          1      100003|
   11  x[3,3]      NL          0           0          1      100000|
   12  x[3,4]      NU          1           0          1       < eps
   13  x[4,1]      NL          0           0          1      100012|
   14  x[4,2]      NL          0           0          1      100007|
   15  x[4,3]      NL          0           0          1      100004|
   16  x[4,4]      NL          0           0          1      100000|

Karush-Kuhn-Tucker optimality conditions:

KKT.PE: max.abs.err = 0.00e+00 on row 0
        max.rel.err = 0.00e+00 on row 0
        High quality

KKT.PB: max.abs.err = 0.00e+00 on row 0
        max.rel.err = 0.00e+00 on row 0
        High quality

KKT.DE: max.abs.err = 0.00e+00 on column 0
        max.rel.err = 0.00e+00 on column 0
        High quality

KKT.DB: max.abs.err = 0.00e+00 on row 0
        max.rel.err = 0.00e+00 on row 0
        High quality

End of output
```

We can use the same model file, Listing 4.4, for different shortest path problems. For example, for the shortest path problem as shown in Figure 1.1, which is an LP problem, we modify only the input file, as shown in **sp-gen2.dat**, Listing 4.8. No modification of the model file is required.

Listing 4.8: Input file: sp-gen2.dat

```
/* sp-gen2.dat */

param p := 1;
param q := 6;
param N := 6;

param : E : cost :=
1 1 100000
1 2 3
1 3 5
1 4 9
1 5 100000
1 6 100000
2 1 100000
2 2 100000
2 3 4
2 4 100000
2 5 4
2 6 100000
3 1 100000
3 2 100000
3 3 100000
3 4 100000
3 5 100000
3 6 10
```

```
26 │ 4 1 1000000
27 │ 4 2 100000
28 │ 4 3 6
29 │ 4 4 100000
30 │ 4 5 100000
31 │ 4 6 14
32 │ 5 1 100000
33 │ 5 2 100000
34 │ 5 3 100000
35 │ 5 4 100000
36 │ 5 5 100000
37 │ 5 6 6
38 │ 6 1 100000
39 │ 6 2 100000
40 │ 6 3 100000
41 │ 6 4 100000
42 │ 6 5 100000
43 │ 6 6 100000
44 │ ;
45 │ end;
```

## 4.1.2   Dijkstra's algorithm

The shortest path from a source node to a destination node can be obtained by Dijkstra's algorithm [1]. It is a more effective way of finding a solution in networks that have non-negative link cost values, compared with the way to solve an LP problem. In Dijkstra's algorithm, nodes in the network are divided into three types of nodes, which are a *visiting* node, a *visited* node, and an *unvisited* node. At the initial stage, the source node is set as a visiting node, and other nodes are set as unvisited nodes. Set the distance to zero for the source node and to infinity ($\infty$) for all other nodes. The distance between nodes adjacent to the visiting node are computed by adding the cost of the link between the visiting node and each adjacent node to the distance of the current node. Next, select the next node to visit. Repeat the process until every node in the network becomes a visited node. This process yields the shortest paths from the source node to all nodes in the network.

The procedure of Dijkstra's algorithm is expressed as follows:

- Step 1: Set distances from a source node to every node, except the source node, to $\infty$, and set the distance of the source node to 0.

- Step 2: Mark the source node as a visiting node, and other nodes as unvisited nodes.

- Step 3: For the visiting node, compute the distance to unvisited adjacent nodes by adding the cost of the link between the current node and each adjacent node to the distance of the current node. If the distance is less than the previously recorded distance, update the distance. For an unvisited node whose distance is updated, the path from the source node via the previous hop node is recorded. The visiting node becomes an visited node. Note that, if a node is an visited node, the distance will not be updated and it is guaranteed to be a component of the shortest path tree.

- Step 4: Choose an unvisited node whose distance is shortest among unvisited nodes and set it to the visiting node.

- Step 5: If all nodes in the network are marked as visited nodes, the process is finished. Otherwise, repeat the same process from step 3.

Figure 4.3 explains how Dijkstra's algorithm finds the shortest path from node 1 to node 6, step by step. $D(i)$ denotes the distance from node 1 to node $i$. Figure 4.3(a) shows the network and link costs. As Step 1, the distance of node 1 is set to 0, and those of other nodes are set to $\infty$, as shown in Figure 4.3(b).

Mark node 1 as a visiting node and mark the other nodes as unvisited nodes, as illustrated in Figure 4.3(c) (step 2). In Figure 4.3, a current node and visited nodes are colored in black, and unvisited nodes are colored in white. Node 1 has three unvisited adjacent nodes, which are nodes 2, 3, and 4. Their distances, which were set to $\infty$, are updated. The distances of node 2, node 3, and node 4 are updated to $D(2) = 2$, $D(3) = 1$, and $D(4) = 5$, respectively (step 3).

In Figure 4.3(d), node 3 is selected because its distance from node 1 is the shortest among the unvisited nodes. Node 1 is marked as a visited node, and node 3 is updated from an unvisited node to the visiting node. Therefore, the shortest path on $1 \rightarrow 3$ is determined and will not be updated (step 4).

Consider three unvisited adjacent nodes, which are nodes 2, 4, and 5, from the visiting node, node 3 (step 3). The distance on path $1 \rightarrow 3 \rightarrow 2$ is 3. However, the previously computed (recorded) distance, $D(2) = 2$, for path $1 \rightarrow 2$, is smaller than $D(2) = 3$ for path $1 \rightarrow 3 \rightarrow 2$. Therefore, we keep $D(2) = 2$ with path $1 \rightarrow 2$. The distance on path $1 \rightarrow 3 \rightarrow 4$ is 4. This value is smaller than the recorded distance, $D(4) = 5$. Therefore, $D(4)$ is updated as $D(4) = 4$ for path $1 \rightarrow 3 \rightarrow 4$. The distance on path $1 \rightarrow 3 \rightarrow 5$ is 2. It is smaller than the recorded distance, $D(5) = \infty$. Therefore, $D(5)$ is updated as $D(5) = 2$. After computing the distances on all the unvisited adjacent nodes, node 3 is marked as a visited node.

Figure 4.3(e), $D(2) = 2$ and $D(5) = 2$ are the smallest values among distances of unvisited nodes. Either node 2 or node 5 can be selected as a visiting node. Let us consider that node 2 is selected as a visiting node. The distance on path $1 \rightarrow 2 \rightarrow 4$ is 5. However, the recorded distance, $D(4) = 4$, from path $1 \rightarrow 3 \rightarrow 4$, is less than that of path $1 \rightarrow 2 \rightarrow 4$. Therefore, $D(4) = 4$ from path $1 \rightarrow 3 \rightarrow 4$ is kept (Step 3). Node 2 becomes a visited node.

In Figure 4.3(f), $D(5) = 2$ is the smallest value among distances of unvisited nodes, so node 5 is selected as a visiting node, Step 4. Node 5 has two unvisited adjacent nodes, which are nodes 4 and 6. The distance on path $1 \rightarrow 3 \rightarrow 5 \rightarrow 4$ is 3. This number is smaller than the previous recorded

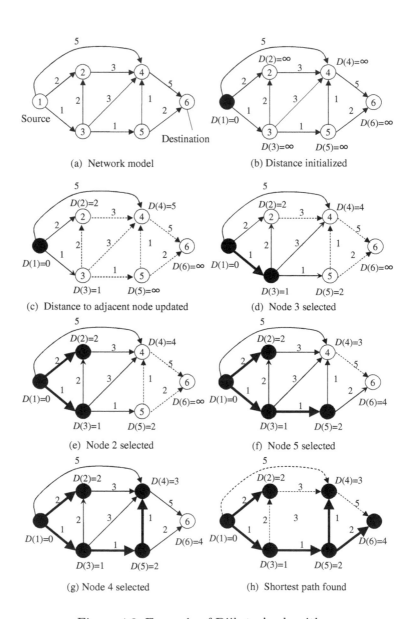

Figure 4.3: Example of Dijkstra's algorithm.

distance, $D(4) = 4$. Therefore, $D(4) = 3$ is overwritten. The distance on path $1 \rightarrow 3 \rightarrow 5 \rightarrow 6$ is 4. We replace $D(6) = \infty$ by $D(6) = 4$, as step 3. Node 5 then becomes a visited node.

In Figure 4.3(g), node 4 is selected as a visiting node (step 4). It has an unvisited adjacent node, node 6. The distance on path $1 \rightarrow 3 \rightarrow 5 \rightarrow 4 \rightarrow 6$ is 8, which is larger than the recorded distance, $D(6) = 4$. Therefore, $D(6) =$ for path $1 \rightarrow 3 \rightarrow 5 \rightarrow 6$ is kept. Node 4 then becomes a visited node. We repeat the process for node 6. Finally, every node in the network becomes a visited node, and the process stops (step 5). Therefore, the shortest path from node 1 to node 6 is found to be $1 \rightarrow 3 \rightarrow 5 \rightarrow 6$, and the distance is 4.

### 4.1.3 Bellman-Ford algorithm

As Dijkstra's algorithm can only be applied to networks that have non-negative link costs, the Bellman-Ford algorithm was introduced to rectify this shortcoming [2]. Note that the correct answer cannot be obtained if there is a negative loop in the network. In a negative loop, the summation of unlimited distance becomes smaller with each iteration. Therefore, the shortest path can-not be determined.

In Dijkstra's algorithm, a node with the shortest distance is selected as a visiting node among unvisited nodes, and the distances of the adjacent nodes from the visiting node are updated or not. Then, the visiting node becomes a visited node, and a new visiting node is found again among the unvisited nodes. Once a node becomes a visited node, the distance is guaranteed to be the shortest one. In the Bellman-Ford algorithm, a node is not categorized by *visited* or not. The distance of every node can be changed up to $N - 1$ times to obtain the shortest path, where $N$ is the number of nodes in the network.

The procedure of the Bellman-Ford algorithm is expressed as follows:

- Step 1: Set distances from the source node to every node, except the source node, to $\infty$, and set the distance for the source node to 0.

- Step 2: At each node, select an adjacent node so that the summation of the distance of the adjacent node and the cost of the link from the adjacent node to its own node is minimized. Set the selected adjacent node to be the previous hop node.

- Step 3: The process finishes if step 2 is repeated $N-1$ times. Otherwise, repeat step 2.

Figure 4.4 shows an example of the Bellman-Ford algorithm. $D(i)$ is denoted as the distance from node 1 to node $i$. $P(i)$ is denoted as a previous hop node of node $i$. The network consists of six nodes, $N = 6$, as shown in Figure 4.4(a). We want to find the shortest path to every node, from node 1. Figure 4.4(b), the distance of node 1 is set to 0 while those of the others are set to $\infty$ (step 1).

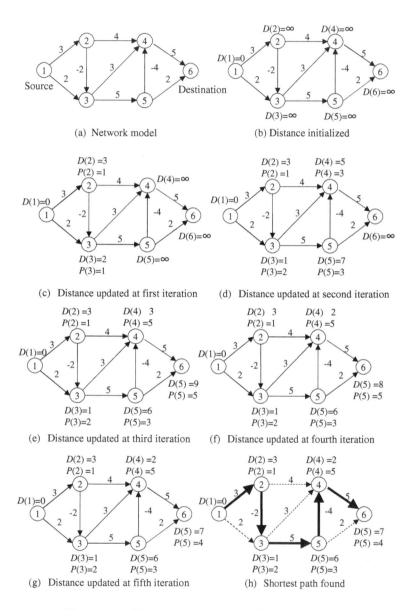

Figure 4.4: Example of Bellman-Ford algorithm.

Figure 4.4(c), distances are updated at the first iteration. The distances, from node 1, to adjacent nodes of all $D(i)$s are computed. Here, adjacent nodes of node 1 with $D(i) \neq \infty$ are node 2 and node 3. Therefore, we compute the distances only for node 2 and node 3. Because node 2 is an adjacent node of node 1, $D(2) = D(1) + 3 = 3 < \infty$ is updated. The previous hop node of node 2 is $P(2) = 1$. Because node 3 is another adjacent node of node 1, $D(3) = D(1) + 2 = 2 < \infty$ is updated. The previous hop node of node 3 is $P(3) = 1$.

Figure 4.4(d) shows distance updates in the second iteration. Nodes with $D(i) \neq \infty$ are node 1, node 2, and node 3. Nodes whose distances will be changed are adjacent nodes of node 1, node 2, and node 3. These are node 2, node 3, node 4, and node 5. An adjacent node is selected so that the summation of the distance of the adjacent node and the cost of the link from the adjacent node to its own node is minimized. As a result, the updated distances are $D(3) = 1$, $D(4) = 5$, and $D(5) = 7$. $P(i)$s become $P(3) = 2$, $P(4) = 3$, and $P(5) = 3$. In the case of negative distances, while $D(3) < D(2)$ is satisfied in Figure 4.4(c), $D(3)$ is updated as shown in Figure 4.4(d). This differs from Dijkstra's algorithm, which accepts only non-negative distances.

The process is repeated until $N - 1 = 5$ times, as shown in Figure 4.4(e), Figure 4.4(f), and Figure 4.4(g). The shortest paths from node 1 to every node are obtained, as shown in Figure 4.4(h).

## 4.2  Max flow problem

### 4.2.1  Linear programming problem

The max flow problem involves finding the traffic flows that maximize traffic volume transmitted from a source node to a destination node, under the constraint that the traffic volume passing through any link cannot exceed that link capacity. Figure 4.5 shows a network model of the max flow problem. The value on each link indicates the link capacity.

The network is represented by directed graph $G(V, E)$, where $V$ is the set of vertexes (nodes) and $E$ is the set of edges (links). A link from node $i$ to node

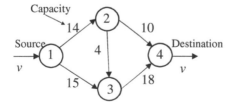

Figure 4.5: Network model of max flow problem. Number on each link represents link capacity.

$j$ is expressed by $(i,j) \in E$. $c_{ij}$ is the link capacity of $(i,j)$. $x_{ij}^{pq}$, where $0 \leq x_{ij}^{pq} \leq c_{ij}$, is the traffic volume from node $p \in V$ to node $q \in V$ routed through $(i,j) \in E$. As the network model in Figure 4.5, the optimization problem that maximizes traffic volume $v$ from source node $p \in V$ to destination node $q \in V$ is expressed as follows:

$$\text{Objective} \quad \max \quad v \tag{4.9a}$$

$$\text{Constraints} \quad x_{12} + x_{13} = v \tag{4.9b}$$

$$x_{12} - x_{23} - x_{24} = 0 \tag{4.9c}$$

$$x_{13} + x_{23} - x_{34} = 0 \tag{4.9d}$$

$$0 \leq x_{12} \leq 14 \tag{4.9e}$$

$$0 \leq x_{13} \leq 15 \tag{4.9f}$$

$$0 \leq x_{23} \leq 4 \tag{4.9g}$$

$$0 \leq x_{24} \leq 10 \tag{4.9h}$$

$$0 \leq x_{34} \leq 18 \tag{4.9i}$$

The decision variables are $v$, $x_{12}$, $x_{13}$, $x_{23}$, $x_{24}$, and $x_{34}$. Eq. (4.9a) represents the objective function that maximizes the traffic volume, $v$ from node 1 to node 4. Eqs. (4.9b)–(4.9i) are the constraints. Eqs. (4.9b)–(4.9d) indicate flow conservation. Eq. (4.9b) is the constraint to maintain flows at the source node, node 1. The outgoing traffic volume from node 1, $x_{12} + x_{13}$, is equal to $v$. Eq. (4.9c) is a constraint to maintain flows at an intermediate, node 2. The incoming traffic volume to node 2, $x_{12}$, is equal to the outgoing traffic volume from node 2, $x_{23} + x_{24}$. Eq. (4.9d) is a constraint to maintain flows at an intermediate node, node 3. The incoming traffic volume to node 3, $x_{13} + x_{23}$, is equal to the outgoing traffic volume from node 3, $x_{34}$. The constraint to maintain flows at the destination node, node 4, $x_{24} + x_{34} = v$, is obtained using Eqs. (4.9b)–(4.9d). Therefore, $x_{24} + x_{34} = v$ is always guaranteed if Eqs. (4.9b)–(4.9d) are satisfied. Eqs. (4.9e)–(4.9i) show the ranges of $x_{ij}$. Each traffic volume passing through $(i,j)$ is less than or equal to the link capacity of $(i,j)$.

Let us solve the LP problem presented in Eqs. (4.9a)–(4.9i) by using GLPK. The model file for Eqs. (4.9a)–(4.9i) is shown in Listing 4.9.

Listing 4.9: Model file: mf-ex1.mod

```
1    /* mf-ex1.mod */
2
3    /* Decision variables */
4    var v ;
5    var x12 <=14, >=0 ;
6    var x13 <=15, >=0 ;
7    var x23 <=4, >=0 ;
8    var x24 <=10, >=0 ;
9    var x34 <=18, >=0 ;
10
11   /* Objective function */
```

```
maximize TRAFFIC: v ;

/* Constraints */
s.t. NODE1: x12 + x13 = v ;
s.t. NODE2: x12 - x23 - x24 = 0 ;
s.t. NODE3: x13 + x23 - x34 = 0 ;

end ;
```

After the program is run using 'glpsol', GLPK displays the process, as shown in Listing 4.10. The output file, 'mf-ex1.out', is obtained, as shown in Listing 4.11. The optimum solution is $v = 28$, $x_{12} = 14$, $x_{13} = 14$, $x_{23} = 4$, $x_{24} = 10$, and $x_{34} = 18$. In other words, there are three routes for the maximum flows. Route 1 is $(1 \rightarrow 2 \rightarrow 4)$ and has traffic volume $v_1 = 10$. Route 2 is $(1 \rightarrow 2 \rightarrow 3 \rightarrow 4)$ and has traffic volume $v_2 = 4$. Route 3 is $(1 \rightarrow 3 \rightarrow 4)$ and has traffic volume $v_3 = 14$. The total traffic volume is $v = v_1 + v_2 + v_3 = 28$.

Listing 4.10: Model file: mf-ex1.txt

```
$ glpsol -m mf-ex1.mod -o mf-ex1.out
GLPSOL: GLPK LP/MIP Solver, v4.45
Parameter(s) specified in the command line:
 -m mf-ex1.mod -o mf-ex1.out
Reading model section from mf-ex1.mod...
20 lines were read
Generating TRAFFIC...
Generating NODE1...
Generating NODE2...
Generating NODE3...
Model has been successfully generated
GLPK Simplex Optimizer, v4.45
4 rows, 6 columns, 10 non-zeros
Preprocessing...
1 row, 2 columns, 2 non-zeros
Scaling...
 A: min|aij| =  1.000e+00  max|aij| =  1.000e+00  ratio =  1.000e+00
Problem data seem to be well scaled
Constructing initial basis...
Size of triangular part = 1
*     0: obj =   1.000000000e+01  infeas =  0.000e+00 (0)
*     2: obj =   2.800000000e+01  infeas =  0.000e+00 (0)
OPTIMAL SOLUTION FOUND
Time used:   0.0 secs
Memory used: 0.1 Mb (107670 bytes)
Writing basic solution to 'mf-ex1.out'...
```

Listing 4.11: Model file: mf-ex1.out

```
Problem:    mf
Rows:       4
Columns:    6
Non-zeros:  10
Status:     OPTIMAL
Objective:  TRAFFIC = 28 (MAXimum)
```

| No. | Row name | St | Activity | Lower bound | Upper bound | Marginal |
|-----|----------|-----|----------|-------------|-------------|----------|
| 1 | TRAFFIC | B | 28 | | | |
| 2 | NODE1 | NS | 0 | -0 | | = |

```
-1
```

```
12  |     3 NODE2        NS           0            -0          =
    | 1
13  |     4 NODE3        NS           0            -0          =
    | 1
14  |
15  |  No. Column name  St   Activity      Lower bound   Upper bound      Marginal
16  | ----- ------------ --  ------------- ------------- ------------- -------------
17  |    1 v             B         28
18  |    2 x12           B         14            0             14
19  |    3 x13           B         14            0             15
20  |    4 x23           NU         4            0              4        < eps
21  |    5 x24           NU        10            0             10            1
22  |    6 x34           NU        18            0             18            1
23  |
24  | Karush-Kuhn-Tucker optimality conditions:
25  |
26  | KKT.PE:  max.abs.err = 0.00e+00 on row 0
27  |          max.rel.err = 0.00e+00 on row 0
28  |          High quality
29  |
30  | KKT.PB:  max.abs.err = 0.00e+00 on row 0
31  |          max.rel.err = 0.00e+00 on row 0
32  |          High quality
33  |
34  | KKT.DE:  max.abs.err = 0.00e+00 on column 0
35  |          max.rel.err = 0.00e+00 on column 0
36  |          High quality
37  |
38  | KKT.DB:  max.abs.err = 0.00e+00 on row 0
39  |          max.rel.err = 0.00e+00 on row 0
40  |          High quality
41  |
42  | End of output
```

The general formulation of the max flow problem is presented as follows:

$$\text{Objective} \qquad \max \quad v \tag{4.10a}$$

$$\text{Constraints} \qquad \sum_{j:(i,j)\in E} x_{ij} - \sum_{j:(j,i)\in E} x_{ji} = v, \quad \text{if } i = p \tag{4.10b}$$

$$\sum_{j:(i,j)\in E} x_{ij} - \sum_{j:(j,i)\in E} x_{ji} = 0, \quad \forall i \neq p, q \in V \tag{4.10c}$$

$$0 \leq x_{ij} \leq c_{ij}, \quad \forall (i,j) \in E \tag{4.10d}$$

The decision variables are $v$ and $x_{ij}$. Eq. (4.10a) represents the objective function that maximizes the traffic volume $v$ from source node $p$ to destination node $q$. Eqs. (4.10b)–(4.10d) shows the constraints. Eq. (4.10b) maintains flows at source node $p$. The outgoing traffic volume from the source node $p$ is equal to $v$, $\sum_{j:(i,j)\in E} x_{ij} - \sum_{j:(j,i)\in E} x_{ji}$. Eq. (4.10c) maintains flows at intermediate nodes, $i$, where $i \neq p, q$. The outgoing traffic volume from node $i$, $\sum_{j:(i,j)\in E} x_{ij}$, is equal to the incoming traffic volume to node $i$, $\sum_{j:(j,i)\in E} x_{ji}$, as shown in Eq. (4.10c). Eq. (4.10d) is the range of $x_{ij}$, which must not exceed $c_{ij}$.

Eqs. (4.10a)–(4.10d) are separately written in a model file as shown in Listing 4.12, and in an input file for the network in Figure 4.5 as shown in Listing 4.5.

Listing 4.12: Model file: mf-gen.mod

```
/* mf-gen.mod */

param N integer, >0 ;
param p integer, >0 ;
param q integer, >0 ;

set V := 1..N ;
set E within {V,V} ;

var TRAFFIC >= 0 ;

param capa{E} ;

var x{E} >= 0 ;
maximize FLOW: TRAFFIC ;
s.t. INTERNAL{i in V: i != p && i != q && p != q }:
                sum{j in V} (x[i,j]) - sum{j in V}(x[j,i]) = 0 ;
s.t. SOURCE{i in V: i = p && p != q}:
                sum{j in V} (x[i,j]) - sum{j in V}(x[j,i]) = TRAFFIC ;
s.t. CAPACITY{(i,j) in E}: x[i,j] <= capa[i,j];
end ;
```

Listing 4.13: Input file: mf-gen1.dat

```
/* mf-gen1.dat */

param p := 1;
param q := 4;
param N := 4;

param : E : capa :=
1 1 0
1 2 14
1 3 15
1 4 0
2 1 0
2 2 0
2 3 4
2 4 10
3 1 0
3 2 0
3 3 0
3 4 18
4 1 0
4 2 0
4 3 0
4 4 0
;
end;
```

Lines 3–5 of input file 'mf-gen1.dat' in Listing 4.13 define parameters $p$, $q$, and $N$. Lines 7–23 define the capacity of each link. In the case that two nodes (e.g., $(1,1)$ or $(1,4)$) have no lonk connecting them, the link capacity is set to 0. The input file of the network in Figure 1.3 is written in 'mf-gen2.dat'; see Listing 4.14. Using the same model file, shown in Listing 4.12, we can obtain the solution, shown in Figure 1.4.

Listing 4.14: Model file: mf-gen2.dat

```
/* mf-gen2.dat */

```

```
 3  param p := 1;
 4  param q := 6;
 5  param N := 6;
 6
 7  param : E : capa :=
 8  1 1 0
 9  1 2 25
10  1 3 100
11  1 4 70
12  1 5 0
13  1 6 0
14  2 1 0
15  2 2 0
16  2 3 30
17  2 4 0
18  2 5 15
19  2 6 0
20  3 1 0
21  3 2 0
22  3 3 0
23  3 4 0
24  3 5 0
25  3 6 200
26  4 1 0
27  4 2 0
28  4 3 60
29  4 4 0
30  4 5 0
31  4 6 30
32  5 1 0
33  5 2 0
34  5 3 0
35  5 4 0
36  5 5 0
37  5 6 150
38  6 1 0
39  6 2 0
40  6 3 0
41  6 4 0
42  6 5 0
43  6 6 0
44  ;
45  end;
```

## 4.2.2   Ford-Fulkerson algorithm

The Ford-Fulkerson algorithm is another approach to solve the max flow problem [3]. The idea of the Ford-Fulkerson algorithm is as follows. Consider a network whose link capacities are given. First, a path that links source and destination nodes is selected and the largest possible traffic flow is assigned under the constraint that the traffic volume passing through each link does not exceed the link capacity. The resulting path is called the augmenting path. The network with residual capabilities after the setting of the augmenting path is considered. This is called the residual network. For the residual network, an additional traffic flow is assigned if possible and it is added to the current flows. The residual network is updated considering the current flows. This process is repeated until no other additional traffic flow can be assigned. After the process is completed, the assigned traffic flow is the sum of the flows on the augmenting paths.

The procedure of the Ford-Fulkerson algorithm is expressed as follows:

- Step 1: Set traffic volumes of every flow to 0.

- Step 2: Create a residual network based on the current flows.

- Step 3: For the residual network defined in Step 2, if there is any path that transmits any traffic from the source node to the destination node, select it. Assign the largest possible traffic flow to the selected path under the constraint that the traffic volume passing through each link does not exceed the link capacity. The assigned flow is added to the current flows, and step 2 is reentered. Otherwise, go to Step 4.

- Step 4: Traffic volume on the current flows in the network becomes the solution of the max flow problem. The process is finished.

Figure 4.6 shows an example of the Ford-Fulkerson algorithm. Figure 4.6(a) is a network model. There are six nodes, node 1 to node 6. We want to find the max flow from the source node, node 1, to the destination node, node 6. This network model is the same as the network model in Figure 1.3.

Figure 4.6(b) shows that a flow is added on the network in Figure 4.6(a). We select path $1 \to 2 \to 5 \to 6$ among the possibilities as an augmenting path at this time. The capacity on this path is limited by link $(2,5)$, because it has the smallest capacity on the path. The maximum traffic volume that can be sent through this path is 15. After the first iteration, the residual network becomes the network in Figure 4.6(b). The capacities of links $(1,2)$, $(2,5)$, and $(5,6)$ are decreased by 15. We then add opposite-directed links $(6,5)$, $(5,2)$, and $(2,1)$ with the value of traffic flow of 15. In other words, the capacity of link $(i, j)$, $Q$, is decreased by traffic flow $p$. Therefore, the residual capacity on $(i, j)$ becomes $Q - p$. We decrease the capacity by adding the opposite-directed link $(j, i)$ with the traffic volume of $p$. Then, a new residual network is created based on the augmenting path of $1 \to 2 \to 5 \to 6$.

Figure 4.6(c) shows that a flow is added to the network in Figure 4.6(b) as the second additional flow. We select path $1 \to 2 \to 3 \to 6$ among the possibilities as an augmenting path at this time. The maximum traffic volume that can be sent through this path is 10, since it is limited by the capacity on $(1, 2) = 10$. At this step, the residual network becomes the network in Figure 4.6(c). We repeat the process for the third additional flow. Path $1 \to 3 \to 6$ is selected. The maximum flow on this path is 100. Path $1 \to 4 \to 3 \to 6$ is selected for the fourth additional flow. The maximum flow on this path is 60. Path $1 \to 4 \to 6$ is selected for the fifth additional flow. The maximum flow on this path is 10. These are shown in Figure 4.6(d), Figure 4.6(e), and Figure 4.6(f).

We are not able to place an additional flow on the residual network, as shown in Figure 4.6(f). The sum of the volumes of current traffic flows that

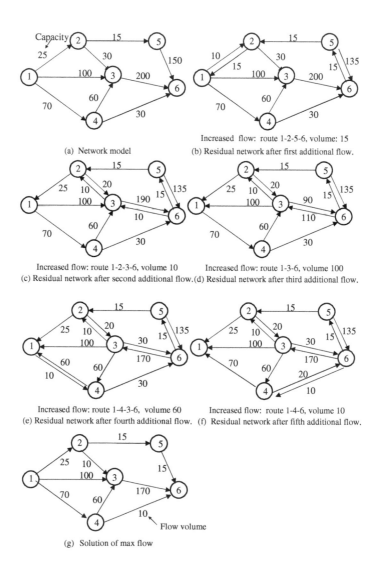

(a) Network model

Increased flow: route 1-2-5-6, volume: 15
(b) Residual network after first additional flow.

Increased flow: route 1-2-3-6, volume 10
(c) Residual network after second additional flow.

Increased flow: route 1-3-6, volume 100
(d) Residual network after third additional flow.

Increased flow: route 1-4-3-6, volume 60
(e) Residual network after fourth additional flow.

Increased flow: route 1-4-6, volume 10
(f) Residual network after fifth additional flow.

(g) Solution of max flow

Figure 4.6: Example of Ford-Fulkerson algorithm.

are directed from node 6 to node node 1 becomes the volume of the max flow that we want to obtain, as shown in Figure 4.6(f). Figure 4.6(g) shows the solution of the max flow problem; the volume of the max flow is 195.

### 4.2.3 Max flow and minimum cut

A subset of nodes that includes source node and excludes destination node is called a cut. There may be several cuts on a network. Figure 4.7 shows only four examples of cuts between the source node, node 1, and the destination node, node 6, for the network in Figure 4.6(a). Other cuts are possible. The total value of link capacities on the cut is called the capacity of the cut. The relationship between the traffic volume and the capacity of the cut is as follows.

**Theorem 4.2.1** *The traffic volume of the max flow is equal to the minimum capacity of the cut.*

In other words, we can find the minimum value of the capacity of the cut to obtain the solution of the max flow. Figure 4.7(a) shows that the minimum cut is 195, which is the same as the max flow solution. Theorem 4.2.1 indicates that the max flow of a network is limited by the minimum capacity of the cut (the bottleneck of the network).

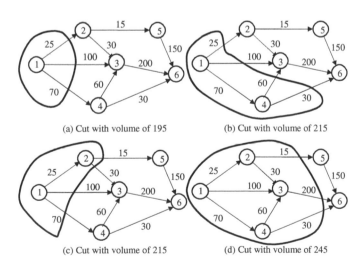

Figure 4.7: Example of max flow and minimum cut.

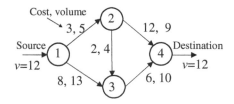

Figure 4.8: Network model of minimum-cost flow problem. Number on each link represents link cost and link capacity.

## 4.3  Minimum-cost flow problem

### 4.3.1  Linear programming problem

The minimum-cost flow problem is to find the traffic flows whose cost is minimized so as to satisfy a traffic demand from a source node to a destination node, under the constraint that the traffic volume passing through each link cannot exceed the link capacity. Figure 4.8 shows a network model of the minimum-cost flow problem. Values on each link represent link cost and link capacity. The required cost for flows each link is defined as the link cost $\times$ the traffic volume passing through the link. We find paths and traffic volumes that achieve the minimum cost.

The network is represented by directed graph $G(V, E)$, where $V$ is the set of vertices (nodes) and $E$ is the set of links. A link from node $i$ to node $j$ is expressed by $(i, j) \in E$. $c_{ij}$ is the link capacity of $(i, j)$. $x_{ij}^{pq}$, where $0 \le x_{ij}^{pq} \le c_{ij}$, is the traffic volume from node $p \in V$ to node $q \in V$ routed through $(i, j) \in E$. Given the network model in Figure 4.5, the minimum-cost flow problem minimizes the total cost required to transmit traffic volume $v$ from source node $p \in V$ and is formulated as follows.

$$
\begin{array}{lll}
\text{Objective} & \min \quad 3x_{12} + 8x_{13} + 2x_{23} + 12x_{24} + 6x_{34} & \text{(4.11a)} \\
\text{Constraints} & x_{12} + x_{13} = 12 & \text{(4.11b)} \\
& x_{12} - x_{23} - x_{24} = 0 & \text{(4.11c)} \\
& x_{13} + x_{23} - x_{34} = 0 & \text{(4.11d)} \\
& 0 \le x_{12} \le 5 & \text{(4.11e)} \\
& 0 \le x_{13} \le 13 & \text{(4.11f)} \\
& 0 \le x_{23} \le 4 & \text{(4.11g)} \\
& 0 \le x_{24} \le 9 & \text{(4.11h)} \\
& 0 \le x_{34} \le 10 & \text{(4.11i)}
\end{array}
$$

The decision variables are $x_{12}$, $x_{13}$, $x_{23}$, $x_{24}$, and $x_{34}$. Eq. (4.11a) indicates the objective function that minimizes the cost to send the demanded traffic volume, $v = 12$, from node 1 to node 4. Eqs. (4.11b)–(4.11i) show the constraints.

Eqs. (4.11b)–(4.11d) represent flow conservation. Eq. (4.11b) maintains the condition of flows at the source node, node 1. The incoming traffic volume to node 1, $x_{12} + x_{13}$, is 12. Eq. (4.11c) maintains the condition of flows at an intermediate node, node 2. The incoming traffic volume , $x_{12}$, and outgoing traffic volume, $x_{23} + x_{24}$, at node 2 are equal. Eq. (4.11d) maintains the condition of flows at an intermediate node, node 3. The incoming traffic volume, $x_{13} + x_{23}$, and the outgoing traffic volume, $x_{34}$, at node 3 are equal. The condition of flows at the destination node, node 4, $x_{24} + x_{34} = 12$, is obtained using Eqs. (4.11b)–(4.11d). Therefore, $x_{24} + x_{34} = 12$ is always guaranteed if Eqs. (4.11b)–(4.11d) are satisfied. Eqs. (4.11e)–(4.11i) show the ranges of $x_{ij}$. Each traffic volume passing through $(i, j)$ is less than or equal to the link capacity of $(i, j)$.

Let us solve the LP problem presented in Eqs. (4.11a)–(4.11i) using GLPK. A model file for Eqs. (4.11a)–(4.11i) is shown in Listing 4.15.

Listing 4.15: Model file: mcf-ex1.mod

```
1   /* mcf-ex1.mod */
2
3   /* Decision variables */
4   var x12 <=5, >=0 ;
5   var x13 <=13, >=0 ;
6   var x23 <=4, >=0 ;
7   var x24 <=9, >=0 ;
8   var x34 <=10, >=0 ;
9
10  /* Objective function */
11  minimize COSTFLOW: 3*x12 + 8*x13 + 2*x23 + 12*x24 + 6*x34 ;
12
13  /* Constraints */
14  s.t. NODE1: x12 + x13 = 12 ;
15  s.t. NODE2: x12 - x23 - x24 = 0 ;
16  s.t. NODE3: x13 + x23 - x34 = 0 ;
17
18  end ;
```

After the program is run by 'glpsol', we obtain $x_{12} = 5$, $x_{13} = 7$, $x_{23} = 3$, $x_{24} = 2$, and $x_{34} = 10$. These three routes are the solution. Route 1 is $(1 \rightarrow 2 \rightarrow 4)$, and its traffic volume is $v_1 = 2$. Route 2 is $(1 \rightarrow 2 \rightarrow 3 \rightarrow 4)$, and its traffic volume is $v_2 = 3$. Route 3 is $(1 \rightarrow 3 \rightarrow 4)$, and its traffic volume is $v_3 = 7$. The total traffic volume is $v = v_1 + v_2 + v_3 = 12$. The total required cost of these traffic flows is 161.

The general formulation of the minimum-cost flow problem is as follows. Traffic volume from node $p$ to node $q$ is $v$. The link cost of $(i, j)$ is defined as $d_{(ij)}$. The link capacity of $(i, j)$ is defined as $c_{ij}$.

$$\text{Objective} \quad \min \quad \sum_{(i,j) \in E} d_{ij} x_{ij} \quad (4.12a)$$

$$\text{Constraints} \quad \sum_{j:(i,j)\in E} x_{ij} - \sum_{j:(j,i)\in E} x_{ji} = v, \quad \text{if } i = p \qquad (4.12b)$$

$$\sum_{j:(i,j)\in E} x_{ij} - \sum_{j:(j,i)\in E} x_{ji} = 0, \quad \forall i \neq p, q \in V \quad (4.12c)$$

$$0 \leq x_{ij} \leq c_{ij}, \quad \forall (i,j) \in E \qquad (4.12d)$$

$x_{ij}$s are decision variables. Eq. (4.12a) indicates the objective function that minimizes the cost to send the demanded traffic volume, $v$, from node $p$ to node $q$. Eqs. (4.12b)–(4.12d) show the constraints. Eqs. (4.12b) and (4.12c) maintain flow conservation. Eq. (4.12b) maintains flows at the source node, node $p$. The outgoing traffic volume from node $p$, $\sum_{j:(i,j)\in E} x_{ij} - \sum_{j:(j,i)\in E} x_{ji}$, is equal to the traffic volume $v$. Eq. (4.12c) maintains flows at intermediate nodes, $i$, where $i \neq p, q$. The outgoing traffic volume, $\sum_{j:(i,j)\in E} x_{ij}$, and the incoming traffic volume, $\sum_{j:(j,i)\in E} x_{ji}$, are equal at each intermediate node. Eq. (4.12d) is the range of $x_{ij}$. Each traffic volume passing through $(i,j)$ is less than or equal to the link capacity of $(i,j)$.

Eqs. (4.12a)–(4.12d) are written, respectively, in a model file, shown in Listing 4.16, and in an input file for the network in Figure 4.8 as shown in Listing 4.17.

Listing 4.16: Model file: mcf-gen.mod

```
1   /* mcf-gen.mod */
2
3   param N integer, >0 ;
4   param p integer, >0 ;
5   param q integer, >0 ;
6   param TRAFFIC, >= 0 ;
7
8   set V := 1..N ;
9   set E within {V,V} ;
10  set EM within E ;
11  param capa{E} ;
12  param cost{EM} ;
13
14  var x{E} >= 0;
15  minimize FLOW_COST: sum{i in V} (sum{j in V} (cost[i,j]*x[i,j] ) ) ;
16  s.t. INTERNAL{i in V: i != p && i != q && p != q }:
17          sum{j in V} (x[i,j]) - sum{j in V}(x[j,i]) = 0 ;
18  s.t. SOURCE{i in V: i = p && p != q}:
19          sum{j in V} (x[i,j]) - sum{j in V}(x[j,i]) = TRAFFIC ;
20  s.t. CAPACITY{(i,j) in E}: x[i,j] <= capa[i,j] ;
21
22  end ;
```

Listing 4.17: Input file: mcf-gen1.dat

```
1   /* mcf-gen1.dat */
2
3   param p := 1 ;
4   param q := 4 ;
5   param N := 4 ;
6   param TRAFFIC := 12 ;
```

```
param : E : capa :=
1 1 0
1 2 5
1 3 13
1 4 0
2 1 0
2 2 0
2 3 4
2 4 9
3 1 0
3 2 0
3 3 0
3 4 10
4 1 0
4 2 0
4 3 0
4 4 0
;
param : EM : cost :=
1 1 100000
1 2 3
1 3 8
1 4 100000
2 1 100000
2 2 100000
2 3 2
2 4 12
3 1 100000
3 2 100000
3 3 100000
3 4 6
4 1 100000
4 2 100000
4 3 100000
4 4 100000
;
end;
```

Lines 3–6 of the input file 'mf-gen1.dat', as in Listing 4.17, define parameters $p$, $q$, $N$, and $v$. Lines 8–24 define the capacity of each link.

In the case that two nodes (e.g., $(1,1)$ or $(1,4)$) have no link, the link capacity is set to 0. Lines 26–42 define the cost of each link. In the case of no link between nodes, the cost is set to a high value. It is set to 10000 in this data file.

An input file for the network in Figure 1.5 is shown by 'mcf-gen2.dat'; see Listing 4.18. Using the same model file as shown in Listing 4.16, we can obtain the solution as shown in Figure 1.6.

Listing 4.18: Input file: mcf-gen2.dat

```
/* mcf-gen2.dat */

param p := 1 ;
param q := 6 ;
param N := 6 ;
param TRAFFIC := 180 ;

param : E : capa :=
1 1 0
1 2 25
1 3 100
```

```
12 │ 1  4  70
13 │ 1  5  0
14 │ 1  6  0
15 │ 2  1  0
16 │ 2  2  0
17 │ 2  3  30
18 │ 2  4  0
19 │ 2  5  15
20 │ 2  6  0
21 │ 3  1  0
22 │ 3  2  0
23 │ 3  3  0
24 │ 3  4  0
25 │ 3  5  0
26 │ 3  6  200
27 │ 4  1  0
28 │ 4  2  0
29 │ 4  3  60
30 │ 4  4  0
31 │ 4  5  0
32 │ 4  6  30
33 │ 5  1  0
34 │ 5  2  0
35 │ 5  3  0
36 │ 5  4  0
37 │ 5  5  0
38 │ 5  6  150
39 │ 6  1  0
40 │ 6  2  0
41 │ 6  3  0
42 │ 6  4  0
43 │ 6  5  0
44 │ 6  6  0
45 │ ;
46 │ param : EM : cost :=
47 │ 1  1  100000
48 │ 1  2  3
49 │ 1  3  5
50 │ 1  4  9
51 │ 1  5  100000
52 │ 1  6  100000
53 │ 2  1  100000
54 │ 2  2  100000
55 │ 2  3  4
56 │ 2  4  100000
57 │ 2  5  4
58 │ 2  6  100000
59 │ 3  1  100000
60 │ 3  2  100000
61 │ 3  3  100000
62 │ 3  4  100000
63 │ 3  5  100000
64 │ 3  6  10
65 │ 4  1  100000
66 │ 4  2  100000
67 │ 4  3  6
68 │ 4  4  100000
69 │ 4  5  100000
70 │ 4  6  14
71 │ 5  1  100000
72 │ 5  2  100000
73 │ 5  3  100000
74 │ 5  4  100000
75 │ 5  5  100000
76 │ 5  6  6
77 │ 6  1  100000
78 │ 6  2  100000
```

```
6 3 100000
6 4 100000
6 5 100000
6 6 100000
;
end;
```

## 4.3.2 Cycle-canceling algorithm

This subsection describes another approach to solving the minimum-cost flow problem, the cycle-canceling algorithm. Also called Klein's algorithm, the concept of the cycle-canceling algorithm is as follows. First, check if there is any feasible flow (solution) that can carry the required traffic demand from a source node to a destination node without considering the cost in the network whose capacities are given. If there is a feasible flow, consider the residual capacities of the network after the traffic flow is assigned. This network is called the residual network. In the residual network, after checking if there is a loop, or a cycle, with negative cost, the optimality of current flow assignment is judged. If there is any negative loop, the current flow is not considered the minimum cost flow because it is possible that the cost could be reduced. The process of flow assignment is repeated until no negative loop exists. When the process finishes, the assigned traffic flow is guaranteed as the minimum cost flow.

The procedure of the cycle-canceling algorithm is as follows:

- Step 1: Check if there is any feasible traffic flow to send the required traffic demand from a source node to a destination node, without consideration of the cost. If there is no feasible traffic flow, the algorithm finishes.

- Step 2: Create a residual network that has residual capacities after the traffic flow is assigned.

- Step 3: For the residual network, check if there is any negative loop; the current flow is not considered the minimum cost flow. Go to Step 4. Otherwise, go to Step 5.

- Step 4: Reduce the cost by injecting traffic flow into the negative loop. The negative loop is then canceled. Then, go to Step 2.

- Step 5: The cost of the traffic flows on the network becomes the minimum. The algorithm is finished.

Figures 4.9–4.11 provide an example of the cycle-canceling algorithm. First, we consider the max flow problem from node 1 to node 4, as in Step 1. Figure 4.9(a) shows the network model considering link costs and link capacities, which is the same as the network in Figure 4.8(a). To determine whether there is a feasible flow, we create an auxiliary network as in Figure 4.9(b),

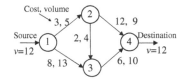

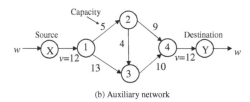

(a) Network for minimum-cost flow problem

(b) Auxiliary network

Figure 4.9: Network for minimum-cost flow problem and auxiliary network.

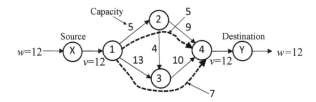

Figure 4.10: Initial flow assignment.

where node $X$ and node $Y$ are newly defined. Consider the network in Figure 4.9(b) and set the capacities of links from node $X$ to node 1 (the original source node), and from node 4 to node $Y$ (the original destination node) to $v$, which is equal to the traffic demand for the original minimum-cost flow problem.

As the objective of Step 1 is to determine whether there is a feasible flow, we do not consider the cost. We solve the max flow problem from the auxiliary network in Figure 4.9(b). Set $w$ to find the maximum traffic volume that can be sent from node $X$ to node $Y$. If $w = v$, there is a feasible solution for the minimum-cost flow problem. If $w < v$, there is no feasible solution. Note that it is impossible for the case of $w > v$ to occur as the capacities of the added links are $v$. Figure 4.10, the solution of the max flow problem is $w = v = 12$. This is an initial flow assignment. There are two traffic flows for the initial flow assignment. The first traffic volume on flow $1 \rightarrow 2 \rightarrow 4$ is 5. The second traffic volume on flow $1 \rightarrow 3 \rightarrow 4$ is 7. This initial flow assignment does not have to be the minimum-cost flow.

In Step 2, create a residual network considering the initial flow assignment. The residual network for the initial flow assignment is shown in Figure 4.11(a).

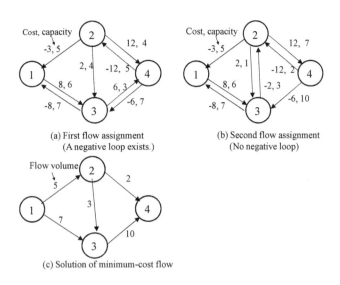

(a) First flow assignment
(A negative loop exists.)

(b) Second flow assignment
(No negative loop)

(c) Solution of minimum-cost flow

Figure 4.11: Example of cycle-canceling algorithm.

The residual network is created by following, for the most part, the method described in Section 4.2.2. For path $1 \to 2 \to 4$ with the traffic volume of 5, the capacities of links $(1, 2)$ and $(2, 4)$ are decreased by 5; $(4, 2)$ and $(2, 1)$, which take the opposite direction on the path, are granted the capacity of 5. The link costs in the opposite direction of the path become $-12$ and $-3$, respectively. For path $1 \to 3 \to 4$ with the traffic volume of 7, the capacities of links $(1, 3)$ and $(3, 4)$ are decreased by 7; $(4, 3)$ and $(3, 1)$, which take the opposite direction on the path, are granted the capacity of 7. The link costs on the opposite direction of the path become $-6$ and $-8$, respectively.

In Step 3, for the residual network in Figure 4.11(a), check if there is a negative loop or not. As total cost of loop $2 \to 3 \to 4 \to 2$ is $2+6+(-12) = -4$, the loop forms a negative loop. As long as there is a negative loop, if we inject a traffic flow into the negative loop, the cost is decreased. Therefore, we inject the maximum traffic volume of 3 into loop $2 \to 3 \to 4 \to 2$. The cost is decreased. Repeat step 2 and create a residual network based on the current flow assignment, as shown in Figure 4.11(b).

There is no negative loop in Figure 4.11(b) and so we are unable to decrease the cost. Therefore, the assigned flow on the network becomes the minimum-cost flow, as shown in Figure 4.11(c).

The cycle-canceling algorithm is based on the following theorem.

**Theorem 4.3.1** *A feasible traffic flow assignment is a solution of the minimum-cost flow problem, if the residual network of the assignment contains no negative loop.*

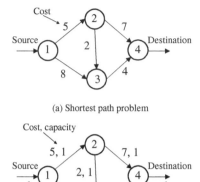

(a) Shortest path problem

(b) Minimum-cost flow problem

Figure 4.12: Shortest path problem and minimum-cost flow problem.

## 4.4    Relationship among three problems

The shortest path problem and the max flow problem are special cases of the minimum-cost flow problem.

The shortest path problem can be solved by the minimum-cost flow problem, as shown in Figure 4.12. However, we consider only link cost in the minimum-cost flow problem. The capacity of each link is set to 1. The traffic demand is assumed to be 1. The solution of this problem becomes a solution of the shortest path problem.

The max flow problem can be solved by the minimum-cost flow problem, as shown in Figure 4.13. In the minimum-cost flow problem, we add a direct connection from a source node to a destination node to the network for the max flow problem. The cost of the link is set to a non-negative value (e.g., 1), and the link capacity is set to a large value (e.g., 1000). In addition, set the traffic demand to a large value (e.g., 1000). For each link in the original network for the max flow problem, keep the link capacities as they are and set each cost to 0. If we solve the minimum-cost flow problem, the flows that are sent through the original network, which does not include for the added link, provide the solution of the max flow problem.

### Exercise 4.1

Find the shortest path from node 1 to node 6 in the network in Figure 4.14. In addition, find the shortest path from node 2 to node 6.

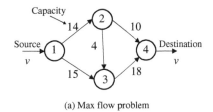

(a) Max flow problem

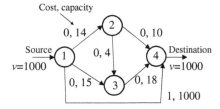

(b) Minimum-cost flow problem

Figure 4.13: Max flow problem and minimum-cost flow problem.

## Exercise 4.2

If link $(4, 5)$ of the network in Figure 4.14 fails, find the shortest path from node 1 to node 6. In addition, find the shortest path from node 2 to node 6.

## Exercise 4.3

Find the minimum-cost flow from node 1 to node 6 in the network in Figure 4.14.

## Exercise 4.4

Find the max flow from node 1 to node 6 in the network in Figure 4.15.

## Exercise 4.5

Find the minimum-cost flow from node 1 to node 4 in the network in Figure 4.16.

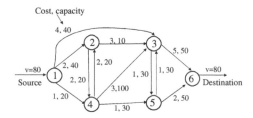

Figure 4.14: Network for shortest path and minimum-cost flow problems.

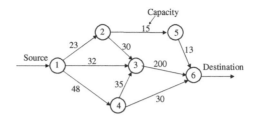

Figure 4.15: Network for max flow problem.

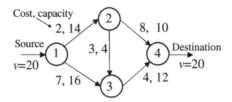

Figure 4.16: Network for minimum-cost flow problem.

# Bibliography

[1] E.W. Dijkstra, "A note on two problems in connexion with graphs," *Numerische Mathematik* 1, pp. 269–271, 1959.

[2] R. Bellman, "On a Routing Problem," *Quarterly of Applied Mathematics*, vol. 16, no. 1, pp. 87–90, 1958.

[3] L. R. Ford, Jr. and D.R. Fulkerson, *Flows in Networks*, Princeton University Press, Princeton, NJ, 1962.

# Chapter 5

# Disjoint path routing

This chapter presents several problems on finding disjoint paths for reliable communications. First, the basic problem of finding a set of disjoint routes whose total cost is minimized, which is called a MIN-SUM problem, is considered. Several approaches, which include integer linear programming (ILP), a disjoint shortest pair algorithm, and the Suurballe algorithm, are introduced to solve the problem. Second, the MIN-SUM problem in a network with shared risk link groups (SRLGs) and its solutions are presented. Third, the MIN-SUM problem in a multiple-cost network and its solutions are introduced.

## 5.1 Basic disjoint path problem

### 5.1.1 Integer linear programming problem

With optical fiber bandwidth and node capacity increasing explosively, a break in a fiber span or node failure can cause a huge amount of damage to customers. Therefore, network providers should design survivable networks that minimize the communication loss. Disjoint path routing enhances the survivability of a network. Several disjoint paths, which are routed without sharing the same links or nodes, must be set between source and destination nodes.

A set of node disjoint paths does not share any common node with any other, while a set of link disjoint paths does not share any common link with any other. A set of link disjoint paths is a subset of node disjoint paths. Let us consider a set of *link* disjoint paths to simplify the discussion. Link disjoint paths are simply called disjoint paths in Sections 5.1 and 5.2, while node disjoint paths are considered in Section 5.3.

Consider the problem of finding a set of $K$ disjoint paths from a source node to a destination node. In the network model in Figure 5.1, a set of two disjoint paths from node 1 to node 6 is searched for, where $K = 2$. A simple algorithm is considered to deal with the problem as follows. The first path is selected as the shortest path of $1 \rightarrow 2 \rightarrow 3 \rightarrow 6$. To find the second path, links used in the first path is deleted as the second path is not allowed to use them. A network for the second path search is shown in Figure 5.2. However, no path is found from node 1 to node 6 in Figure 5.2. If a path of $1 \to 4 \to 3 \to 6$ is selected as the first path, the second path is found as $1 \to 2 \to 5 \to 6$. Therefore, the algorithm in which the shortest path is selected as the first path does not always give any feasible solution for this problem, even if a feasible solution exists.

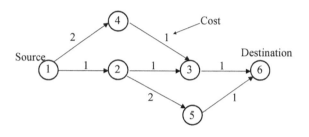

Figure 5.1: Network model with link costs.

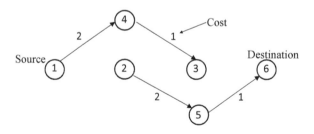

Figure 5.2: Network after the first path is selected.

Next, consider the problem of finding a set of $K$ disjoint paths from a source to a destination node so that the total costs of the paths can be minimized. The problem is called the MIN-SUM problem. The network is represented by directed graph $G(V, E)$, where $V$ is the set of vertices (nodes) and $E$ is the set of links. A link from node $i$ to node $j$ is expressed by $(i, j) \in E$. $x_{ij}^k$ is the portion of the traffic on path $k \in M$ from node $p \in V$ to node $q \in V$ routed through $(i, j) \in E$, where $x_{ij}^k$ takes a binary value, 0 or 1, and

$M = \{1, 2, \cdots, K\}$. If path $k$ is routed through $(i, j)$, $x_{ij}^k = 1$. Otherwise, $x_{ij}^k = 0$. $d_{ij}$ is the link cost of $(i, j)$.

The MIN-SUM problem, finding a set of $K$ disjoint paths from a source to a destination node, is formulated as the following ILP problem:

$$\text{Objective} \quad \min \sum_{k \in M} \sum_{(i,j) \in E} d_{ij} x_{ij}^k \tag{5.1a}$$

$$\text{Constraints} \quad \sum_{j:(i,j) \in E} x_{ij}^k - \sum_{j:(j,i) \in E} x_{ji}^k = 1, \quad \forall k \in M, \text{if } i = p \tag{5.1b}$$

$$\sum_{j:(i,j) \in E} x_{ij}^k - \sum_{j:(j,i) \in E} x_{ji}^k = 0, \quad \forall k, i(\neq p, q) \in V \tag{5.1c}$$

$$x_{ij}^k + x_{ij}^{k'} \leq 1, \quad \forall k, k'(k \neq k') \in M, (i, j) \in E \tag{5.1d}$$

$$x_{ij}^k = \{0, 1\}, \quad \forall k \in M, (i, j) \in E \tag{5.1e}$$

$x_{ij}$ and $d_{ij}$ are the decision variable and the link cost of $(i, j)$, respectively. Eq. (5.1a) is the objective function that minimizes the total costs of $K$ paths. Eqs. (5.1b)–(5.1e) are the constraints. Eqs. (5.1b) and (5.1c) express the conditions of flow conservation. Eq. (5.1b) maintains the flows at the source node, node $p$. The different between the incoming traffic volume and the outgoing traffic volume, $\sum_{j:(i,j) \in E} x_{ij} - \sum_{j:(j,i) \in E} x_{ji}$, is 1. Here, the outgoing traffic volume at node $p$ is 1. Eq. (5.1c) maintains flows at intermediate node $i$, where $i \neq p, q$. The outgoing traffic volume at node $i$, $\sum_{j:(i,j) \in E} x_{ij}$, is equal to the incoming traffic volume at node $i$, $\sum_{j:(j,i) \in E} x_{ji}$. Eq. (5.1d) indicates that different paths must not share any common link. Eq. (5.1e) expresses the range of $x_{ij}$.

Eqs. (5.1a)–(5.1e) are separately written in the model file as shown in Listing 4.4 and input file that represents a network in Figure 5.2 as shown in Listing 5.2.

Listing 5.1: Model file: djp-gen.mod

```
1  /* djp-gen.mod */
2
3  /* Given parameters */
4  param K integer , >0 ;
5  param N integer , >0 ;
6  param p integer , >0 ;
7  param q integer , >0 ;
8
9  set V := 1..N ;
0  set E within {V,V} ;
1  set M := 1..K ;
2
3  param cost{E};
4
5  /* Decision variables */
6  /* var x{E,M} >=0, <=1, integer ; */
7  var x{E,M} binary ;
8
9  /* Objective function */
20 minimize PATH_COST: sum{k in M} sum{i in V} (sum{j in V}
```

```
21                          (cost[i,j]*x[i,j,k])) ;
22
23   /* Constraints */
24   s.t. SOURCE{i in V, k in M: i = p && p != q}:
25         sum{j in V} (x[i,j,k]) - sum{j in V}(x[j,i,k]) = 1 ;
26   s.t. INTERNAL{i in V, k in M: i != p && i != q && p != q }:
27         sum{j in V} (x[i,j,k]) - sum{j in V}(x[j,i,k]) = 0 ;
28   s.t. DISJOINT{i in V, j in V, k1 in M, k2 in M: k2 !=k1}:
29         x[i,j,k1] + x[i,j,k2] <= 1 ;
30   end;
```

Lines 4, 5, 6, 7, and 13 of the model file in List 5.1 define the types of parameters for the number of paths, $K$, the number of nodes, $N$, source node, $p$, destination node, $q$, and link costs, respectively.

Listing 5.2: Input file: djp-gen1.dat

```
1    /* djp-gen1.dat */
2
3    param K := 2 ;
4    param N := 6 ;
5    param p := 1 ;
6    param q := 6 ;
7
8    param : E : cost :=
9    1 1 100000
10   1 2 1
11   1 3 100000
12   1 4 2
13   1 5 100000
14   1 6 100000
15   2 1 100000
16   2 2 100000
17   2 3 1
18   2 4 100000
19   2 5 2
20   2 6 100000
21   3 1 100000
22   3 2 100000
23   3 3 100000
24   3 4 100000
25   3 5 100000
26   3 6 1
27   4 1 100000
28   4 2 100000
29   4 3 1
30   4 4 100000
31   4 5 100000
32   4 6 100000
33   5 1 100000
34   5 2 100000
35   5 3 100000
36   5 4 100000
37   5 5 100000
38   5 6 1
39   6 1 100000
40   6 2 100000
41   6 3 100000
42   6 4 100000
43   6 5 100000
44   6 6 100000
45   ;
46   end;
```

Lines 3–6 of the input file in Listing 5.2 define the values of parameters $K$, $N$, $p$, and $q$. Lines 8–44 define the link cost of $(i, j)$. To handle the case of two nodes with no link between them, the cost for $(i, j)$ is set to a large enough number that the pair will never be considered in forming the selected paths. In this case, we set the cost to 10000.

After the program is run using 'glpsol', we find two disjoint routes of $1 \rightarrow 4 \rightarrow 3 \rightarrow 6$ and $1 \rightarrow 2 \rightarrow 5 \rightarrow 6$. The minimum value, which is the total cost of the two paths, is 8.

## 5.1.2 Disjoint shortest pair algorithm

The MIN-SUM problem, finding a set of two disjoint paths from a source to a destination node, can be solved by the disjoint shortest pair algorithm [1]. It is a more effective way of finding a solution, compared to solving an ILP problem, presented in Section 5.1.1.

The procedure of the disjoint shortest pair algorithm is expressed as follows:

- Step 1: Find the first shortest path in a given network.

- Step 2: Reverse the directions of links on the first shortest path and make these link costs negative.

- Step 3: Find the second shortest path in the modified network.

- Step 4: A duplicated link that is used in both first and second shortest paths with different directions is deleted.

- Step 5: The remaining links on the first and second shortest paths form two disjoint paths.

Figure 5.3 explains how the disjoint shortest pair algorithm finds two disjoint paths from node 1 to node 6 to minimize the total cost of the paths, step by step. Figure 5.3(a) shows the network and link costs. In Step 1, the first shortest path, $1 \rightarrow 2 \rightarrow 3 \rightarrow 6$, is found as shown in Figure 5.3(b). In Step 2, the directions of links on the found shortest path, which are $(1, 2)$, $(2, 3)$, and $(3, 6)$, are reversed, and these link costs are made negative. The modified network is made as shown in Figure 5.3(c). In step 3, the second shortest path, $1 \rightarrow 4 \rightarrow 3 \rightarrow 2 \rightarrow 5 \rightarrow 6$, is found for the modified network, as shown in Figure 5.3(d). Note that, in the case, there is a link cost that has a negative value; the Bellman-Ford algorithm, not Dijkstra's algorithm, is used to find the shortest path. In Step 4, as shown in Figure 5.3(e), a duplicated link that is used in both the first and second shortest paths with different directions, which is $(2, 3)$, is deleted. In Step 5, as shown in Figure 5.3(f), the remaining links on the first and second shortest paths form two disjoint paths, which are $1 \rightarrow 4 \rightarrow 3 \rightarrow 6$ and $1 \rightarrow 2 \rightarrow 5 \rightarrow 6$.

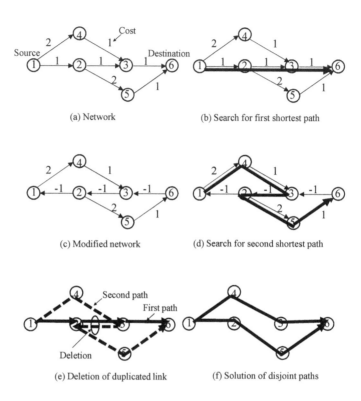

Figure 5.3: Example of disjoint shortest path algorithm.

## 5.1.3  Suurballe's algorithm

In the disjoint shortest pair algorithm, even when a network does not have any negative link cost, the Bellman-Ford algorithm, not Dijkstra's algorithm, must be used, as negative link costs appear in the modified network. Suurballe's algorithm [2, 3] is also an algorithm that can solve the MIN-SUM problem to find a set of two disjoint paths from a source to a destination node. In Suurballe's algorithm, as the modified network does not include any negative link cost, the algorithm is able to use Dijkstra's algorithm, which has lower computation complexity than the Bellman-Ford algorithm.

The procedure of Suurballe's algorithm is expressed as follows:

- Step 1: Find the shortest path tree and the first shortest path in a given network.

- Step 2: Using the shortest path tree and the first shortest path obtained in Step 1, a modified network is constructed in the following way. The link cost of $(i, j)$ in the modified network, denoted as $d'_{ij}$, is given by

$$d'_{ij} = d_{ij} - D(j) + D(i),  \tag{5.2}$$

where $d_{ij}$ is the link cost of $(i, j)$ and $D(i)$ is the total cost from the source node to node $i$ in the given network. In the modified network, the directions of links on the first shortest path are reversed.

- Step 3: Find the second shortest path in the modified network.

- Step 4: A duplicated link that is used in both the first and second shortest paths with different directions is deleted.

- Step 5: The remaining links on the first and second shortest paths form two disjoint paths.

Consider that the given network has non-negative link costs. In Step 2, if $(i, j)$ is on the shortest path tree obtained in Step 1, $d'_{ij} = 0$ by $D(j) = D(i) + d_{ij}$. Otherwise, $d'_{ij} \geq 0$ as $D(j) \geq D(i) + d_{ij}$. As a result, all link costs in the modified network are non-negative.

Figure 5.4 explains how Suurballe's algorithm finds two disjoint paths from node 1 to node 6 to minimize the total cost of the paths, step by step. Figure 5.4(a) shows the network and link costs. Step 1 finds the shortest path tree from node 1 and the first shortest path, $1 \rightarrow 2 \rightarrow 3 \rightarrow 6$, as shown in Figure 5.4(b). Step 2 modifies the network, as shown in Figure 5.4(c). The directions of links on the found shortest path, which are $(1, 2)$, $(2, 3)$, and $(3, 6)$, are reversed, and the link cost of $(i, j)$, $d'_{ij}$, is given by Eq. (5.2). In Step 3, the second shortest path, $1 \rightarrow 4 \rightarrow 3 \rightarrow 2 \rightarrow 5 \rightarrow 6$, is found for the modified network, as shown in Figure 5.4(d). As the modified network does not have any negative link cost, Dijkstra's algorithm is used to find the shortest path. In Step 4, as shown in Figure 5.4(e), the duplicated link that is used in both first and second shortest paths with different directions, which is $(2, 3)$, is deleted. In Step 5, as shown in Figure 5.4(f), the remaining links on the first and second shortest paths form two disjoint paths, which are $1 \rightarrow 4 \rightarrow 3 \rightarrow 6$ and $1 \rightarrow 2 \rightarrow 5 \rightarrow 6$.

## 5.2 Disjoint paths with shared risk link group

### 5.2.1 Shared risk link group (SRLG)

SRLG is defined as a group of links that are affected simultaneously when a network failure occurs. In the network in Figure 5.5, $S(i, j, g)$ indicates the SRLG information, where $S(i, j, g) = 1$ or means that link (i,j) does or does not belong to SRLG $g$, respectively; (5,7) and (3,7) belong to the same SRLG

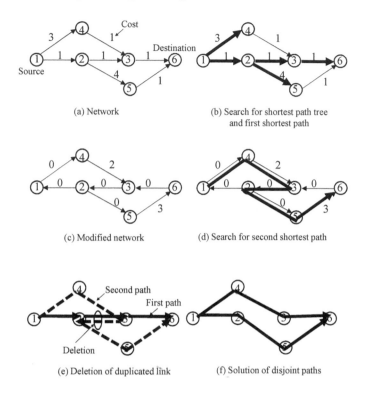

Figure 5.4: Example of Suurballe's algorithm.

group 1. Figure 5.6 shows an SRLG example for (5,7) and (3,7). Both links use the same optical fiber by using wavelengths $\lambda 1$ and $\lambda 2$, respectively. Nodes 3, 5, and 7 are routers, which are switch packets, and an optical crossconnect that links these nodes, which are switch wavelength paths. If the optical fiber between the crossconnect and node 7 is cut, (3,7) and (5,7) are disconnected simultaneously as $\lambda 1$ and $\lambda 2$ are not available. Note that although (5,6) and (2,6) belong to only SRLG 1 in Figure 5.6, a link may generally belong to more than one SRLG. For example, a link could belong to two SRLGs when one SRLG is a optical fiber and the other is a conduit.

Let us find two disjoint paths from node 1 to node 7 to minimize the total cost of the paths. If we search for two disjoint paths without considering SRLG by simply solving the MIN-SUM problem presented in Section 5.1, we obtain the two paths of $1 \rightarrow 2 \rightarrow 3 \rightarrow 7$ and $1 \rightarrow 4 \rightarrow 5 \rightarrow 7$. However, using (3,7) and (5,7) for different disjoint paths yields the risk of simultaneous failure, as they both belong to SRLG 1. Therefore, if we consider SRLG, these paths are not disjoint in terms of SRLG.

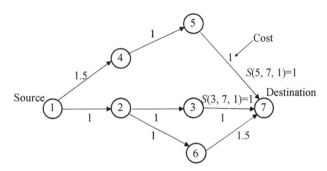

Figure 5.5: Network model with SRLG. If $(i, j)$ belongs to SRLG $g$, $S(i, j, g) = 1$, otherwise $S(i, j, g) = 0$.

## 5.2.2  Integer linear programming

Let us find $K$ disjoint paths from a source to a destination to minimize the total cost of the paths, considering SRLG so that at least one path is available if a single network failure occurs. The network is represented by directed graph $G(V, E)$, where $V$ is the set of vertices (nodes) and $E$ is the set of links. A link from node $i$ to node $j$ is expressed by $(i, j) \in E$. $x_{ij}^k$ is the portion of the traffic on path $k \in M$ from node $p \in V$ to node $q \in V$ routed through $(i, j) \in E$, where $x_{ij}^k$ takes a binary value, 0 or 1, and $M = \{1, 2, \cdots, K\}$. If path $k$ is routed through $(i, j)$, $x_{ij}^k = 1$. Otherwise, $x_{ij}^k = 0$. $d_{ij}$ is the link cost of $(i, j)$.

The MIN-SUM problem to find a set of $K$ disjoint paths from a source to a destination node considering SRLG is formulated as the following ILP problem.

$$\text{Objective} \quad \min \sum_k \sum_{ij} d_{ij} x_{ij}^k \tag{5.3a}$$

$$\text{Constraints} \quad \sum_{j:(i,j)\in E} x_{ij}^k - \sum_{j:(j,i)\in E} x_{ji}^k = 1, \quad \forall k \in M, \text{if } i = p \tag{5.3b}$$

$$\sum_{j:(i,j)\in E} x_{ij}^k - \sum_{j:(j,i)\in E} x_{ji}^k = 0, \quad \forall k \in M, i(\neq p, q) \in V \tag{5.3c}$$

$$x_{ij}^k + x_{ij}^{k'} \leq 1, \quad \forall k, k' \in M, (i, j) \in E \tag{5.3d}$$

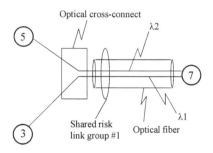

Figure 5.6: Example of SRLG.

$$x_{ij}^k + x_{i'j'}^{k'} + S(i,j,g) + S(i',j',g) \leq 3,$$
$$\forall k, k', (k \neq k') \in M, (i,j), (i',j'), ((i,j) \neq (i',j')) \in E$$

(5.3e)

$$x_{ij}^k = \{0,1\}, \quad \forall k \in M, (i,j) \in E \tag{5.3f}$$

When SRLG is considered, Eq. (5.3e) is added to Eqs. (5.1b)–(5.1e) as a constraint. In Eq. (5.3e), when $(i,j)$ and $(i',j')$, where $(i,j) \neq (i',j')$, are on paths $k$ and $k'$, where $k \neq k'$, respectively, it cannot be allowed that both $S(i,j,g) = 1$ and $S(i',j',g) = 1$. In other words, if $x_{ij}^k + x_{i'j'}^{k'} = 2$, $S(i,j,g) + S(i',j',g) \leq 2$.

Eqs. (5.3b)–(5.3f) are separately written in the model file as shown in Listing 5.3 and the input file that represents the network in Figure 5.5 as shown in Listing 5.4.

Listing 5.3: Model file: djp-s-gen.mod

```
1   /* djp-s-gen.mod */
2
3   /* Given parameters */
4   param K integer, >0 ;
5   param N integer, >0 ;
6   param p integer, >0 ;
7   param q integer, >0 ;
8   param G integer, >0 ;
9
10  set V := 1..N ;
11  set E within {V,V} ;
12  set M := 1..K ;
13  set R := 1..G ;
14  set ER within {E,R} ;
15
```

```
param cost{E} ;
param S{ER} ;

/* Decision variables */
var x{E,M} binary ;

/* Objective function */
minimize PATH_COST: sum{k in M} sum{i in V} (sum{j in V}
                        (cost[i,j]*x[i,j,k])) ;

/* Constraints */
s.t. SOURCE{i in V, k in M: i = p && p != q}:
     sum{j in V} (x[i,j,k]) - sum{j in V}(x[j,i,k]) = 1 ;
s.t. INTERNAL{i in V, k in M: i != p && i != q && p != q }:
     sum{j in V} (x[i,j,k]) - sum{j in V}(x[j,i,k]) = 0 ;
s.t. DSJ{i in V, j in V, k1 in M, k2 in M: k2 !=k1}:
     x[i,j,k1] + x[i,j,k2] <= 1 ;
s.t. SRLG_DSJ{(i1,j1) in E, (i2,j2) in E, g in R, k1 in M, k2 in M:
     k2 != k1 && !(i1=i2 && j2=j2)}:
     x[i1,j1,k1] + x[i2,j2,k2] + S[i1,j1,g] + S[i2,j2,g] <=3 ;
end ;
```

Lines 4–8, 16, and 18 of the model file in List 5.3 define the types of parameters for the number of paths, $K$; the number of nodes, $N$; source node, $p$; destination node, $q$; the number of SRLGs, $G$; link costs, and SRLGs, respectively.

Listing 5.4: Input file: djp-s-gen1.dat

```
/* djp-s-gen1.dat */

param K := 2 ;
param N := 7 ;
param p := 1 ;
param q := 7 ;
param G := 1 ;

param : E : cost :=
1 1 100000
1 2 1
1 3 100000
1 4 1.5
1 5 100000
1 6 100000
1 7 100000
2 1 100000
2 2 100000
2 3 1
2 4 100000
2 5 100000
2 6 1
2 7 100000
3 1 100000
3 2 100000
3 3 100000
3 4 100000
3 5 100000
3 6 100000
3 7 1
4 1 100000
4 2 100000
4 3 100000
4 4 100000
4 5 1
4 6 100000
```

```
37  | 4  7  100000
38  | 5  1  100000
39  | 5  2  100000
40  | 5  3  100000
41  | 5  4  100000
42  | 5  5  100000
43  | 5  6  100000
44  | 5  7  1
45  | 6  1  100000
46  | 6  2  100000
47  | 6  3  100000
48  | 6  4  100000
49  | 6  5  100000
50  | 6  6  100000
51  | 6  7  1.5
52  | 7  1  100000
53  | 7  2  100000
54  | 7  3  100000
55  | 7  4  100000
56  | 7  5  100000
57  | 7  6  100000
58  | 7  7  100000
59  | ;
60  | param : ER : S :=
61  | 1  1  1  0
62  | 1  2  1  0
63  | 1  3  1  0
64  | 1  4  1  0
65  | 1  5  1  0
66  | 1  6  1  0
67  | 1  7  1  0
68  | 2  1  1  0
69  | 2  2  1  0
70  | 2  3  1  0
71  | 2  4  1  0
72  | 2  5  1  0
73  | 2  6  1  0
74  | 2  7  1  0
75  | 3  1  1  0
76  | 3  2  1  0
77  | 3  3  1  0
78  | 3  4  1  0
79  | 3  5  1  0
80  | 3  6  1  0
81  | 3  7  1  1
82  | 4  1  1  0
83  | 4  2  1  0
84  | 4  3  1  0
85  | 4  4  1  0
86  | 4  5  1  0
87  | 4  6  1  0
88  | 4  7  1  0
89  | 5  1  1  0
90  | 5  2  1  0
91  | 5  3  1  0
92  | 5  4  1  0
93  | 5  5  1  0
94  | 5  6  1  0
95  | 5  7  1  1
96  | 6  1  1  0
97  | 6  2  1  0
98  | 6  3  1  0
99  | 6  4  1  0
100 | 6  5  1  0
101 | 6  6  1  0
102 | 6  7  1  0
103 | 7  1  1  0
```

```
7 2 1 0
7 3 1 0
7 4 1 0
7 5 1 0
7 6 1 0
7 7 1 0
;
end;
```

Lines 3–7 of the input file in Listing 5.4 define the values of parameters $K$, $N$, $p$, $q$, and $G$. Lines 9–58 define the link cost of $(i, j)$. To handle the case of two nodes with no link between them, the cost for $(i, j)$ is set to a large enough number that the pair will never be considered in forming the selected paths. In this case, we set the cost to 10000. Lines 60–109 define the SRLGs. $S(3, 7, 1)$ and $S(5, 7, 1)$ are set to 1, and others are set to 0.

After the program is run using 'glpsol', we find two disjoint paths of $1 \rightarrow 2 \rightarrow 6 \rightarrow 7$ and $1 \rightarrow 4 \rightarrow 5 \rightarrow 7$, as shown in Figure 5.7. The minimum value, which is the total costs of the two paths, is 7.

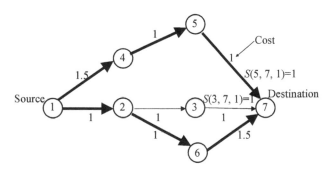

Figure 5.7: Example of disjoint path solution considering SRLG.

### 5.2.3 Weight-SRLG algorithm

When network size becomes large, the complexity of the ILP computation needed to tackle the MIN-SUM problem with SRLG presented in Section 5.2.2 increases and becomes difficult to solve. This section introduces a heuristic algorithm to find disjoint paths with SRLG constraints. It is called the weighted-SRLG algorithm (WSRLG) [4].

A $k$-shortest path algorithm is widely used to find disjoint paths because of its simplicity [1, 5, 6]. Initially, the first path is searched for a given network topology. Node disjoint paths are considered as disjoint paths in this chapter. Next, the links and nodes used by the first path are deleted from the given network topology. Note that deleting a node means that links that are connected to it are also deleted. Then, the second shortest path is searched for in the modified network topology. In the same way, the $k$-shortest disjoint path is searched for. Although the $k$-shortest path algorithm does not find the *maximum* number of disjoint paths, Dunn et al. [7] showed that its results are nearly equal to the max flow solution.

WSRLG treats the total number of SRLG members related to a link as part of the link cost when the $k$-shortest path algorithm is executed. In WS-RLG, a link that has many SRLG members is rarely selected as the shortest path. Oki et al. observed that WSRLG finds more disjoint paths than the conventional $k$-shortest path algorithm [4]. In addition, because WSRLG searches for the weight of the SRLG factor using a modified binary search algorithm while satisfying the required number of disjoint paths between source and destination nodes, it can find cost-effective disjoint paths.

Let us define terminologies used in WSRLG. The network is represented by directed graph $G(V, E)$, where $V$ is the set of vertexes (nodes) and $E$ is the set of links. A link from node $i$ to node $j$ is expressed by $(i, j) \in E$. $d_{ij}$ is the link cost of $(i, j)$ in a given network. $C_{comp}(i, j)$ is the cost that is used in the disjoint path computation. $S(i, j, g)$ indicates the SRLG information, where $S(i, j, g) = 1$ or means that link $(i, j)$ does or does not belong to SRLG $g$, respectively. $\alpha$ is a weight factor for SRLG. $D_{req}(p, q)$ is the required number of disjoint paths from source node $p$ to destination node $q$. $C_{path}(p, q)$ is the sum of costs for all disjoint paths between nodes $p$ and $q$. $K(p, q)$ is the number of obtained disjoint paths between nodes $p$ and $q$. $N_s$ is the number of SRLG groups.

$C_{comp}(i, j)$ is defined by

$$C_{comp}(i, j) = \frac{1 - \alpha}{d_{ij}^{max}} d_{ij} + \frac{\alpha}{SRLG^{max}} \max\{SRLG(i, j), 1\} \tag{5.4}$$

$$SRLG(i, j) = \sum_{g}^{N_s} S(i, j, g) \tag{5.5}$$

$$d_{ij}^{max} = \max_{i,j} d_{ij} \tag{5.6}$$

and

$$SRLG^{max} = \max_{i,j} SRLG(i, j). \tag{5.7}$$

$C_{path}(p, q)$ is expressed by

$$C_{path}(p, q) = \sum_{k=1}^{K(p,q)} \sum_{(i,j) \in \text{path } k} d_{ij}, \tag{5.8}$$

where path $k$ is the $k$th shortest path found by the $k$-shortest path algorithm.

When $SRLG(i, j) = 0$, $\max\{SRLG(i, j), 1\}$ in the second term of Eq.( 5.4) is set to 1 so that the hop count for the link can be considered. Therefore, the value of the second term in Eq.( 5.4) is affected by the hop count even when $(i, j)$ does not belong to any SRLG group, or $SRLG(i, j) = 0$. If $(i, j)$ does not belong to any SRLG, $SRLG(i, j) = 0$. $SRLG^{max}$ determines the sensitivity of $\alpha$. The smaller the value of $SRLG^{max}$, the more sensitive to $\alpha$ the value of $C_{comp}(i, j)$ is.

In WSRLG, $\alpha$ is set to an appropriate value using a modified version of the well-known binary search method [8] so that $C_{path}(p, q)$ can be reduced as much as possible under the condition that $K(p, q)$ is equal to or larger than $D_{req}(p, q)$.

Next, we explain why we modified the conventional binary search method. In our estimate, $K(p, q)$ and $C_{path}(p, q)$ mostly increase with $\alpha$. However, this estimate is not always true. $K(p, q)$ and $C_{path}(p, q)$ does not always increase monotonically with $\alpha$. Therefore, the conventional binary search method may miss an appropriate $\alpha$ that satisfies the required number of disjoint paths. As a result, the $\alpha$ that is finally obtained by the conventional binary search method does not satisfy the required number of disjoint paths. To avoid this problem, the modified method searches for $\alpha$ while remembering the most appropriate candidate in the regular binary search process.

The procedure of WSRLG is expressed below, step by step. As initial values, $\alpha_{min} = 0.0$, $\alpha_{max} = 1.0$, $K_{temp} = \infty$, and $C_{temp} = \infty$ are set. Here, $\epsilon$ is used as a parameter to judge whether the modified binary search method converges. It should be set considering the value of $SRLG^{max}$, which determines the sensitivity of $\alpha$.

Step 1: $\alpha = \dfrac{\alpha_{min} + \alpha_{max}}{2}$.

Step 2: $K(p, q)$ is calculated by the following $k$-shortest path algorithm considering the SRLG constraints.

Step 3: If $K(p, q) \geq D_{req}(p, q)$, then $\alpha_{max} = \alpha$ is set. Otherwise, $\alpha_{min} = \alpha$ is set.

Step 4: If $K(p, q) \geq D_{req}(p, q)$ and $C_{path}(p, q) < C_{temp}$, then $\alpha_{temp} = \alpha$, $K_{temp} = K(p, q)$, and $C_{temp} = C_{path}(p, q)$ are set.

Step 5: If $\alpha_{max} - \alpha_{min} > \epsilon$, go to step 1. Otherwise, go to Step 6.

Step 6: If $K_{temp} \geq D_{req}(p, q)$, then $\alpha = \alpha_{temp}$. A set of disjoint paths obtained with this $\alpha$ is considered a solution. Otherwise, no set of disjoint paths that satisfies the required conditions is found.

The $k$-shortest path algorithm with SRLG is described below. First, we set $k = 1$ as the initial value.

Step 1: The $k$th shortest path between source and destination nodes is searched for based on link cost $C_{comp}(i, j)$. If the path is found, go to Step 2. Otherwise, $K(p, q) = k$ is set and the $k$-shortest algorithm is ended.

Step 2: Delete $(i, j)$ and nodes that are on the $k$th shortest path. For all $g$s, if $S(i, j, g) = 1$, all links $(i', j') = 1$, where $S(i', j', g) = 1$, are also deleted.

Step 3: We set $k = k + 1$ and go to Step 2.

Consider WSRLG applied to the network model in Figure 5.5 to find two disjoint paths from node 1 to node 7. Here, $\alpha$ is set to 1.0, as a special case. WSRLG finds two disjoint paths of $1 \to 2 \to 6 \to 7$ and $1 \to 4 \to 5 \to 7$, which is the same result as that using ILP, as shown in Figure 5.7. $C_{comp}(2, 6)$ and $C_{comp}(5, 6)$ are twice as large as other $C_{comp}(i, j)$. The path of $1 \to 2 \to 6 \to 7$ is selected as the first shortest path first. Then, the path of $1 \to 4 \to 5 \to 7$ is found as the second shortest path. Therefore, WSRLG is able to find more disjoint paths than the conventional $k$-shortest path algorithm.

## 5.3    Disjoint paths in multi-cost networks

### 5.3.1    Multi-cost networks

In the problems presented in Sections 5.1 and 5.2, the cost of each link is considered to be the same for all $k$ paths. Such a network is called a *single-cost* network. On the other hand, the network in which each link can have a different cost for $k$ paths is called a *multi-cost* network. The applications of such a network usually lie in the field of shared backup path protection [9,10]. In this case, the cost of a link for a backup path is often a fraction of that for a working path. A backup path can share the bandwidth with other backup paths, while a working path is unable to share the bandwidth with any other path. As sharing the bandwidth leads to reduce the link cost, the link cost is considered to depend on the level of sharing.

Examples of single-cost and multi-cost networks are shown in Figure 5.8, where a number associated with each link indicates its cost. In a single-cost network, one link cost is given for each link for all three disjoint paths in a single-cost network. Each path cost is defined by the sum of the link costs along the path in the network, as shown in Figure 5.8(a). In a multi-cost network, a different link cost, which is associated with the $k$th path, is given for each in a multi-cost network. The $k$th path cost is defined as the sum of the link costs, which may be different from those for other different paths, along the path in the network, as shown in Figure 5.8(b).

This section considers the MIN-SUM problem, which is a problem to find disjoint paths to minimize the total cost of all paths, in a multi-cost network. This section focuses on *node* disjoint paths. Node disjoint paths are simply called disjoint paths in this section.

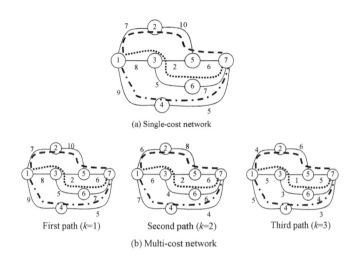

Figure 5.8: Single-cost and multi-cost networks.

## 5.3.2 Integer linear programming problem

Consider the MIM-SUM problem in a multi-cost network to find a set of $K$ disjoint paths from a source to a destination node. The network is represented by directed graph $G(V, E)$, where $V$ is the set of vertices (nodes) and $E$ is the set of edges (links). A link from node $i$ to node $j$ is expressed by $(i, j) \in E$. $x_{ij}^k$ is the portion of the traffic on path $k \in M$ from node $p \in V$ to node $q \in V$ routed through $(i, j) \in E$, where $x_{ij}^k$ takes a binary value, 0 or 1, and $M = \{1, 2, \cdots, K\}$. If path $k$ is routed through $(i, j)$, $x_{ij}^k = 1$. Otherwise, $x_{ij}^k = 0$. $d_{ij}^k$ is the link cost of $(i, j)$ on the $k$th path.

The MIN-SUM problem in a multi-cost network is formulated as an ILP problem in the following:

$$\text{Objective} \quad \min \sum_{k \in M} \sum_{(i,j) \in E} d_{ij}^k x_{ij}^k \tag{5.9a}$$

$$\text{Constraints} \quad \sum_{j:(i,j) \in E} x_{ij}^k - \sum_{j:(j,i) \in E} x_{ji}^k = 1, \quad \forall k \in M, \text{if } i = p, \tag{5.9b}$$

$$\sum_{j:(i,j) \in E} x_{ij}^k - \sum_{j:(j,i) \in E} x_{ji}^k = 0, \quad \forall k, i(\neq p, q) \in V \tag{5.9c}$$

$$x_{ij_1}^1 + \cdots + x_{ij_K}^K \leq 1, \quad \forall i(\neq p), j_1, \cdots, j_K \in V \qquad (5.9\text{d})$$

$$x_{j_1 i}^1 + \cdots + x_{j_K i}^K \leq 1, \quad \forall i(\neq q), j_1, \cdots, j_K \in V \qquad (5.9\text{e})$$

$$x_{ij}^k = \{0, 1\}, \quad \forall k \in M, (i, j) \in E \qquad (5.9\text{f})$$

Eq. (5.9a) is the objective function that minimizes the total costs of $K$ paths in the multi-cost network. Eqs. (5.9b)–(5.9c) maintain flow conservation. To ensure that the paths of each connection do not traverse the common transit nodes, the node disjoint constraints are needed, as shown in Eqs. (5.9d) and (5.9e) Eq. (5.9e) is the binary constraint for the ILP formulation.

As is the same with other ILP problems, when a network size becomes large, the computation complexity of ILP increases and it becomes difficult to solve it [11]. The following subsection introduces two heuristic algorithms, which are called $k$-penalty with auxiliary link costs matrix (KPA) [12] and $k$-penalty with initial link costs matrix (KPI) [13,14].

### 5.3.3    KPA: $k$-penalty with auxiliary link costs matrix

J. Rak presented the KPA algorithm to find $k$-disjoint paths in a multi-cost network in [12]. KPA finds the shortest path as the first path. The links on the shortest path and those connected to the transit nodes on this path are considered as forbidden links for the next disjoint path to be found. Although the $k$-shortest path algorithm, which is applied to a single-cost network, assigns the forbidden links infinitely high cost [1,5,6], KPA gives them finite costs to avoid the trap problem [12]. That is, the next path must pay a penalty for using a forbidden link. Forbidden link cost is increased by the path cost of the previously found path. Link costs are incrementally updated and kept in an auxiliary link cost matrix. If any conflict (i.e. the current path is not disjoint with all previously found paths) occurs, all found paths are deleted. Before starting the process of finding $k$ disjoint paths again, the costs of conflicting links are incrementally increased by the cost of the last found path in the previous iteration. The path cost is computed using the auxiliary link cost matrix.

#### 5.3.3.1    Terminology

The terminology in this section is shown in the following.

$d_r$      Demand to find a set of end-to-end $k$ disjoint paths between a pair of nodes $(s_r, t_r)$

$s_r$      Source node of demand $d_r$

$t_r$      Destination node of demand $d_r$

$i_{max}$      Maximum allowable number of conflicts

$p$      Index of path $1, \cdots, k$

$\eta_p$      $p$th path

| | |
|---|---|
| $a_h$ | $h$th link, where $h = 1, 2, \ldots$ |
| $\xi_h$ | Cost of each link $a_h$ |
| $\xi_h^p$ | Cost of link $a_h$ of the $p$th path |
| $\xi^p$ | Cost of $p$th path that is a sum of $\xi_h^p$ over traversed $a_h$ |
| $\xi_h^{aux}$ | Auxiliary cost of link $a_h$ |
| $\xi_h^{aux,p}$ | Auxiliary cost of link $a_h$ of the $p$th path |
| $\Xi^p$ | Initial matrix of link cost $\xi_h^p$ |
| $\Xi^{aux}$ | Auxiliary matrix of link cost $\xi_h^{aux}$ |
| $\Xi^{aux,p}$ | Auxiliary matrix of link cost $\xi_h^{aux,p}$ |
| $i_c$ | Conflict counter |

### 5.3.3.2 Description

The procedure of KPA is presented in the following:

INPUT:      Demand $d_r$ to find the set of $k$-disjoint paths between a pair of nodes$(s_r, t_r)$.

The initial link costs matrices $\Xi^1, \Xi^2, \ldots, \Xi^k$ (one matrix for each path) of a demand.

The maximum allowable number of conflicts, $i_{max}$.

OUTPUT:     The set of $k$-disjoint paths $\eta_1, \eta_2, \ldots, \eta_k$ — all between a given pair of demand source and destination nodes $(s_r, t_r)$.

The total path cost of $k$-disjoint paths is

$$\xi^{total} = \sum_{p=1}^{k} \sum_{a_h \text{ on path } \eta_p} \xi_h^p. \tag{5.10}$$

Step 1:     Set $i_c = 1$ and $\Xi^{aux,p} = \Xi^p$ for $p = 1, \cdots, k$.

Step 2:     Set $j = 1$.

Step 3:     Set $\Xi^{aux} = \Xi^{aux,j}$.

Step 4:     Consider each path $\eta_i$ from the set of previously found $j - 1$ paths and for each link $a_h$, if $a_h$ is a forbidden link of the path $\eta_i$, then increase the link cost $\xi_h^{aux}$ by path cost $\xi^{aux,i}$ of $\eta_i$ on the network with costs matrix $\Xi^{aux,i}$. That is,

$$\xi_h^{aux} = \xi_h^{aux} + \xi^{aux,i}. \tag{5.11}$$

The path cost is defined by

$$\xi^{aux,i} = \sum_{a_h \text{ on path } \eta_i} \xi_h^{aux,i} \text{for } i = 1, \ldots, j - 1. \tag{5.12}$$

Step 5:   Find the shortest path $\eta_j$ on the network with costs matrix $\Xi^{aux}$.

Step 6:   If $\eta_j$ is disjoint with the previously found $j-1$ paths, then set $j = j + 1$ and go to Step 7.
else

6a) Increase the costs $\xi_h^{aux,1}, \cdots, \xi_h^{aux,k}$ of each conflicting link $a_h$ of $\eta_j$ by path cost $\xi^{aux}$ of $\eta_j$ on the network with cost matrix $\Xi^{aux}$. That is,

$$\xi_h^{aux,p} = \xi_h^{aux,p} + \xi^{aux} \text{ when } p = 1, \cdots, k, \qquad (5.13)$$

where path cost $\xi^{aux}$ is defined by

$$\xi^{aux} = \sum_{a_h \text{ on path } \eta_j} \xi_h^{aux}, \qquad (5.14)$$

then delete the found paths and set $i_c = i_c + 1$.

6b) If $i_c > i_{max}$, then terminate and reject the demand, else go to Step 2.

Step 7:   If $j > k$, then terminate and return the found set of paths, else go to Step 3.

Demand $d_r$ to find $k$ disjoint paths from source node $s_r$ to destination node $t_r$, the link cost matrices for each disjoint path, and the maximum allowable number of conflicts, $i_{max}$, are initially given. KPA outputs the set of $k$ disjoint paths and the total costs of the $k$ disjoint paths. KPA uses the shortest-path-based algorithm. At Step 1, the conflict counter, $i_c$, is set to 1 and the initial cost matrix of the $p$th path, $\Xi^p$, is copied to the auxiliary cost matrix of the $p$th path, $\Xi^{aux,p}$, for all paths, $p = 1, \ldots, k$. $\Xi^p$ is kept to compute the total path cost using Eq. (5.10) after finding $k$ disjoint paths. At Step 2, set $j = 1$ to find the first path. In Step 3, $\Xi^{aux,p}$ is copied to $\Xi_h^{aux}$. Step 4 is skipped if $j = 1$. To find the next paths $\eta_j$ ($j \neq 1$), path $\eta_j$ has to pay a penalty for using one of the forbidden links, that is, links traversed by previously found paths $\eta_i$ (link disjoint), or links corrected to transit nodes used by previously found paths (node disjoint). The cost of the forbidden links is increased by the costs of all $j-1$ previously found paths at Step 4. At Step 5, $\eta_j$ is found as the shortest path on the network with auxiliary cost matrix $\Xi^{aux}$. At Step 6, if $\eta_j$ is disjoint with the $(j-1)$ previously found paths, the index number of path $j$ is increased by one and the process goes to Step 7 to check if the required number of $k$ disjoint paths has been obtained. The process terminates if the number of found disjoint paths has reached the required number, $k$. Otherwise, the process will find the next path by reentering Step 3. If $\eta_j$ is not disjoint (link or node) with the $(j-1)$ previously found paths, a conflict is called, and the costs $\xi_h^{aux,1}, \ldots, \xi_h^{aux,k}$ of each conflicting link $a_h$, the link shared between the previously found $j - 1$ paths and path $\eta_j$, or the link connected to the node shared between previously found $j-1$ paths and path $\eta_j$, is increased by the path cost $\xi^{aux}$ of $\eta_j$, which is computed from auxiliary costs matrix $\Xi^{aux}$ (Step 6a) as shown in Eq. (5.13). After increasing each conflicting link $a_h$, all found paths are deleted and conflict counter, $i_c$, is increased by one. If $i_c$ is greater than the maximum allowable number of conflicts, $i_{max}$, the process is terminated. If $i_c$ is less than $i_{max}$, the process reenters Step 2.

### 5.3.3.3 Example of KPA

KPA is demonstrated with an example in Figure 5.9. The example shows how to find the $k$ disjoint paths in a multi-cost network with $k = 3$ for the demand between node 1 to 7. Figures 5.9(a1), (a2), and (a3) illustrate multi-cost network that has three sets of link costs; one for the first path, $\xi_h^1$; the second path: $\xi_h^2$; and the third path: $\xi_h^3$. KPA starts by setting the auxiliary cost matrix as $\Xi^{aux,p} = \Xi^p$ for $p = 1, 2, 3$ and $i_c = 1$. KPA then considers at the first path $j = 1$ and sets the auxiliary cost matrix $\Xi^{aux} = \Xi^{aux,j}$. The first path $\eta_1$ (1-4-7) is found as the shortest path, shown as Figure 5.9(b1). The costs, $\xi_h^{aux,1}$, for links incident to transit nodes of path $\eta_1$ of the set of link costs $\xi_h^{aux,2}$ are increased by path cost $\xi^{aux,1}$ of path $\eta_1$, which is equal to 14 in the example, as shown in Figure 5.9(b2). Then, path $\eta_2$ (1-3-5-7) is found. To find the third path, the costs of forbidden links of paths $\eta_1$ and $\eta_2$ on the network are increased by path cost $\xi^{aux,1}$ of path $\eta_1$ and path cost $\xi^{aux,2}$ of path $\eta_2$. However, $\eta_3$ (1-4-7), which is not disjoint with $\eta_1$, is found, as shown in Figure 5.9(b3). Costs $\xi_h^{aux,p}$ of links incident to node 4 on $\eta_3$ for all paths, $p = 1, .., k$, are increased by path cost $\xi^{aux}$, as shown in Eq. (5.13). The path cost $\xi^{aux}$ is defined by Eq. (5.14), which is equal to 36 in the example, as shown in Figure 5.9(c1). Next, all found paths are deleted and $i_c$ is increased by one. KPA starts finding $k$ disjoint paths from the beginning again, as shown in Figure 5.9(c1). However, KPA takes time to find the required set of $k$ node-disjoint paths because it avoids the paths with high cost, and this situation leads to overlap with used paths, as shown in Figures 5.9(d3), (e3), (f3), (g3) and (h3). Finally, KPA finds a set of $k = 3$ node-disjoint paths, which are $\eta_1$ (1-3-6-7), $\eta_2$ (1-2-5-7), and $\eta_3$ (1-4-7), at $i_c = 8$, as shown in Figure 5.9(i3).

Leepila et al. found that KPA sometimes fails with a large number of iterations, even though disjoint paths actually exist [13, 14]. With every iteration, or conflict, the link costs in the auxiliary link cost matrix are increased. In order to avoid traversing link with large costs, the algorithm may find a path that overlaps already used paths, including forbidden links from the previous paths. This path overlapping causes the deletion of found paths and restarts the process. It takes time to find $k$ disjoint paths or sometimes they cannot be found at all. This problem must be solved to find $k$ disjoint paths in an efficient manner.

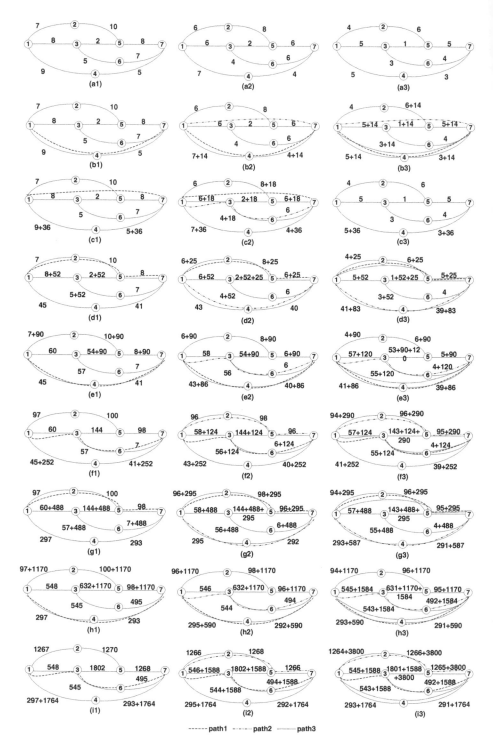

Figure 5.9: Example of KPA for multi-cost network; demand $d_r = (1, 7)$ and $k = 3$. (©2011 IEEE, Ref. [13].)

### 5.3.4 KPI: $k$-penalty with initial link costs matrix

Leepila et al. presented the KPI algorithm to solve the problem of KPA in [13, 14] by extending KPA. KPI uses the same penalty process as KPA, but only the policy of updating $\xi_h^{aux,p}$ is different from KPA (Step 6a). KPI increases the costs $\xi_h^{aux,1}, \ldots, \xi_h^{aux,k}$ of each conflicting link $a_h$ of $\eta_j$ by path cost $\xi^j$ of $\eta_j$ using initial costs matrix $\Xi^j$. Step 6a of KPI is as follows:

6a) Increase the cost $\xi_h^{aux,1}, \ldots, \xi_h^{aux,k}$ of each conflicting link $a_h$ of $\eta_j$ by path cost $\xi^j$ of $\eta_j$ on the network with cost matrix $\Xi^j$. That is,

$$\xi_h^{aux,p} = \xi_h^{aux,p} + \xi^j, \text{ when } p = 1, \ldots, k, \tag{5.15}$$

where path cost $\xi^j$ is defined by

$$\xi^j = \sum_{a_h \text{ on path } \eta_j} \xi_h^j \tag{5.16}$$

and then delete the found paths and set $i_c = i_c + 1$.

#### 5.3.4.1 Example of KPI

We reuse the example in demonstrating KPI; see Figure 5.10. The algorithm starts finding the first path $\eta_1$ (1-4-7) using the shortest-path-based algorithm, Figure 5.10(b1). Before finding path $\eta_2$, the cost $\xi_h^{aux}$ of links incident to transit nodes of path $\eta_1$ are increased by the total cost $\xi^{aux,1}$ of path $\eta_1$, which is equal to 14 in the example, Figure 5.10(b2). After that path $\eta_2$ (1-3-5-7) is found. The cost, $\xi_h^{aux}$, of links incident to transit nodes of paths $\eta_1$ and $\eta_2$ are increased by path cost $\xi^{aux,1}$ of path $\eta_1$ and path cost $\xi^{aux,2}$ of path $\eta_2$, respectively. However, $\eta_3$ (1-4-7), which is not disjoint with $\eta_1$ and has a common transit node, node 4, is found as shown in Figure 5.10(b3). Cost $\xi_h^{aux,p}$ of links incident to node 4 for all paths, $p = 1, ..., k$, are increased by path cost $\xi^3$ computed from the initial link costs of $\eta_3$ defined by Eq. (5.16), which is equal to 8 in Figure 5.10(c1). Next, all the found paths are deleted and $i_c$ is increased by one. The algorithm starts from the beginning, as shown in Figure 5.10(c1). Finally, KPI finds a set of $k = 3$ node-disjoint paths, which are $\eta_1$ (1-2-5-7), $\eta_2$ (1-4-7), and $\eta_3$ (1-3-6-7), as shown in Figure 5.10(d3). Because KPI is more careful in increasing the costs of conflicting links, it can find a set of $k = 3$ node-disjoint paths at the conflict counter $i_c$ value of 3 in the same way as KPA. This example shows that KPI can find a set of $k$ node-disjoint paths faster than KPA.

### 5.3.5 Performance comparison of KPA and KPI

We compare the KPI performance to that of KPA using computer simulations of the U.S. long-distance network and the Italian network; see Figure 5.11 [12].

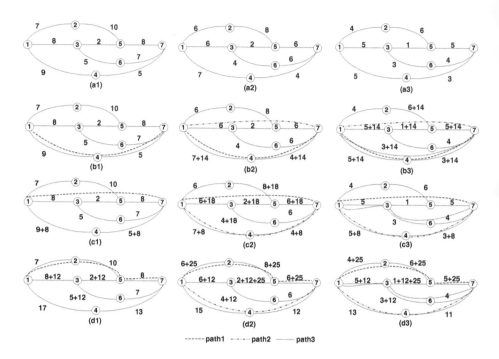

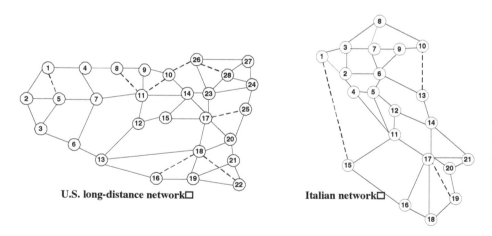

Figure 5.10: Example of KPI for multi-cost network; demand $d_r = (1, 7)$ and $k = 3$. (©2011 IEEE, Ref. [13].)

Figure 5.11: U.S. long-distance network and Italian network.

The required number of disjoint paths was set to $k = 3$. Therefore, additional links, shown as dashed lines in both networks, were needed to keep the degree of each node greater than or equal to 3. The link cost matrices of the multi-cost networks were set to 3 for $k = 3$. One hundred link cost matrices for each corresponding disjoint path were generated uniformly in a random manner in the range of $0 < \xi_h \leq 1$, where $\xi_h$ is the cost of link $a_h$. For both KPI and KPA, we examined the average probability that $k$ disjoint paths were successfully found within a specified maximum allowable number of conflicts, $i_{max}$, over all source and destination node pairs for all generated cost matrices. The probability is defined as the success ratio of finding $k$ disjoint paths.

Figures 5.12(a) and (b) show that KPI finds $k$ disjoint paths faster than the KPI for both Italian network and U.S. long-distance network. Regardless of $i_{max}$, KPI has a higher success ratio than KPA. In addition, KPI yields success ratios of more than 99% with $i_{max} = 10$ for both networks, while KPA does not reach 99% when $i_{max}$ becomes large. In KPA, the conflict path cost defined in Eq. (5.14) is set at too large a value, the conflicting links are always avoided. On the other hand, as KPI defines the conflict path cost according to Eq. (5.16), the conflicting links are appropriately utilized.

The total cost of $k$ disjoint paths, which is defined in Eq. (5.10), is lower with KPI than with KPA. Figure 5.13 compares the normalized total costs of $k$ disjoint paths, normalized by the total path cost of $k$ disjoint paths by Bhandari's algorithm. We used Bhandari's algorithm, which is an algorithm for finding $k$ disjoint paths in single-cost network, to find $k$ disjoint paths using only the link cost matrix for the first path. After $k$ disjoint paths are found by Bhandari's algorithm, the total path cost in a multi-cost network is calculated using Eq. (5.10) with three link costs matrices. The normalized costs for Bhandari's, KPA, and KPI are taken as average values over all source and destination node pairs for all generated cost matrices. The results indicate that KPI yields lower total path cost of than KPA or Bhandari's algorithm for both Italian and U.S. long-distance multi-cost network, as shown in Figure 5.13. Bhandari's algorithm yields the highest path cost among the three algorithm, as it considers only the link cost matrix for the first path to find $k$ disjoint paths. Because, in KPI, the conflicting path cost is suitably estimated and the conflicting links are appropriately utilized, it returns the lowest total path costs.

## Exercise 5.1

Find two disjoint paths from node 1 to node 12 to minimize the total costs of the paths in the network in Figure 5.14.

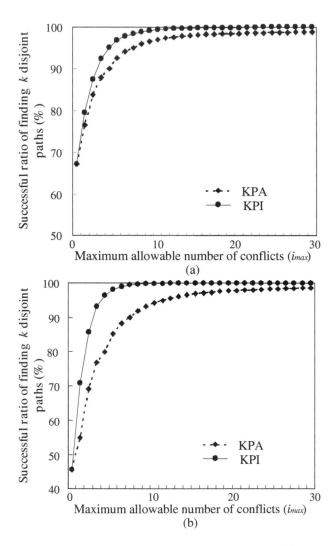

Figure 5.12: Successful ratio of finding $k$ disjoint paths (%) within specified maximum allowable number of conflicts, $i_{max}$ on (a) U.S. long-distance network and (b) Italian network. (©2011 IEEE, Ref. [13].)

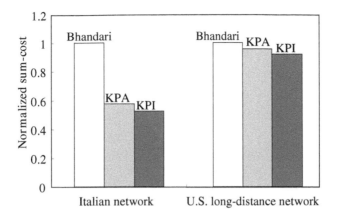

Figure 5.13: Normalized summation of $k$ disjoint paths costs on U.S. long-distance network and Italian network using Bhandari's, KPA, and KPI algorithms. (©2011 IEEE, Ref. [13].)

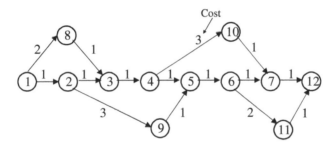

Figure 5.14: Network for disjoint path problem.

# Bibliography

[1] R. Bhandari, *Survivable Networks: Algorithms for Diverse Routing*, 477, Springer 1999.

[2] J.W. Suurballe,, "Disjoint paths in a network," *Networks*, vol. 4, no. 2, pp. 125–145, 1974.

[3] J.W. Suurballe and R.E. Tarjan, "A quick method for finding shortest pairs of disjoint paths," *Networks*, vol. 14, no. 2, pp. 325–336, 1984.

[4] E. Oki, N. Matsuura, K. Shiomoto, and N. Yamanaka, "A disjoint path selection scheme with shared risk link groups in GMPLS networks," *IEEE Commun. Letters*, vol. 6, no. 9, pp. 406–408, Sept. 2002.

[5] E. Oki and N. Yamanaka, "A recursive matrix-calculation method for disjoint path search with hop link number constraints," *IEICE Trans. Commun.*, vol. E78-B, no. 5, pp. 769–774, 1995.

[6] D. Xu, Y. Chen, Y. Xiong, C. Qiao, and X. He, "on the complexity of and algorithms for finding the shortest path with a disjoint counterpart," *IEEE/ACM Trans. on Networking*, vol. 14, no. 1, pp. 147–158, Feb. 2006.

[7] D. A. Dunn, W. D. Grover, and M. H. MacGregor, "Comparison of *k*-shortest paths and maximum flow routing for network facility restoration," *IEEE J. Selct. Areas Commun.*, vol. 12, no. 1, pp. 88–99, Jan. 1994.

[8] A. V. Aho, J. E. Hopcroft, and J. D. Ullman, *The Design and Analysis Computer Algorithms*, Addison-Wesley Series in Computer Science, Boston, 1974.

[9] B.G. Jozsa, D. Orincsay, and A. Kern, "Surviving multiple network failures using shared backup path protection," *Proc. of the 8th IEEE Symp. on Comput. and Commun.*, pp. 1333–1340, July 2003.

[10] J. Tapolcai, P.H. Ho, D. Verchere, T. Cinkler, and A. Haque, "A new shared segment protection method for survivable networks with guaranteed recovery time," *IEEE Trans. on Reliability*, vol. 57, no. 2, pp. 272–282, June. 2008.

[11] M. R. Garey, and D. S. Johnson, *Computers and Intractability: A Guide to the Theory of NP-Completeness*, W. H. Freeman and Company, San Francisco, pp. 217–218, 1979.

[12] J. Rak, "*k*-Penalty: A novel approach to find *k*-disjoint paths with differentiated path costs," *IEEE Commun. Letters*, vol. 14, no. 4, pp. 354–356, Apr. 2010.

[13] R. Leepila, E. Oki, and N. Kishi, "Scheme to find *k* disjoint paths in multi-cost networks," *IEEE ICC 2011*, June 2011.

[14] R. Leepila, E. Oki, and N. Kishi, "Survivable Multi-Cost Networks with *k* Disjoint Paths," *Cyber Journals: Multidisciplinary Journals in Science and Technology, Journal of Selected Areas in Telecommunications (JSAT)*, pp. 26–33, July 2011.

# Chapter 6

# Optical wavelength-routed network

Wavelength division multiplexing (WDM) is a technology to increase the transmission capacity in an optical fiber, where multiple wavelengths carry data simultaneously. In addition to the effect of increasing the transmission capacity, WDM is also useful for wavelength-based switching, which enables us to set an optical path, which is routed on several fibers by connecting each wavelength per fiber through optical crossconnect(s). A network that is formed by several optical paths is called an optical path network, where each wavelength is related to a different optical path destination. Different paths accommodated in each fiber must use different wavelengths. This raises the problem of how to assign wavelengths to the paths while minimizing the number of wavelengths required. The problem is called a wavelength assignment problem. This chapter introduces wavelength assignment problems in an optical wavelength-routed network.

## 6.1 Wavelength assignment problem

In an Internet Protocol (IP) network, a router receives a packet, analyzes the destination address of the packet header electrically, and transmits it to the next hop. In an optical path network, an optical path transits several optical crossconnects, each of which switches the optical signal carried on each wavelength to the intended direction. Switching in an optical crossconnect can be performed without electrical processing.

An example of an optical network is shown in Figure 6.1. Two optical paths are established. One is routed on $A \to X \to y \to D$ with wavelength $\lambda 1$; the other is routed on $B \to X \to y \to C$ with wavelength $\lambda 2$. Crossconnect $X$ receives optical signals from different fibers and injects them into one fiber with different wavelengths. Crossconnect $Y$ receives one optical signal with different wavelengths in the fiber and injects them into two fibers, where each intended direction is determined using wavelength information.

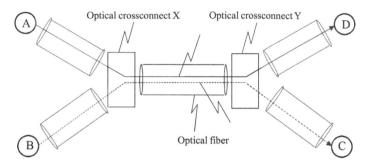

Figure 6.1: Optical path network.

If optical paths are to be established, we need to consider the problem of how to assign a wavelength to each optical path. This problem is called the wavelength assignment problem. Assume that the routes of these optical paths are already determined. In this problem, different optical paths passing through the same fiber must use different wavelengths. To deploy network systems in an optical path network, it is desirable to minimize the number of wavelengths used in the network. Therefore, in the wavelength assignment problem, we consider the objective function of minimizing the number of wavelengths.

Let us consider a network with five optical path requests, as shown in Figure 6.2. Figure 6.3 shows an example, where wavelengths are assigned sequentially in increasing order of the path indices. First, $\lambda_1$ is assigned to optical path 1. Next, consider the assignment of a wavelength to optical path 2. As optical path 2 shares optical fiber with optical path 1, $\lambda_2$, which is a different wavelength from that of optical path 1, is assigned to optical path 2. Third, consider wavelength assignment for optical path 3. As optical path 3 shares optical fiber with optical paths 1 and 2, $\lambda_3$, which is a different wavelength from those of optical paths 1 and 2, is assigned to optical path 3. Fourth, consider wavelength assignment for optical path 4. As optical path 4 shares optical fiber with optical path 3, but does not share any optical fiber with optical paths 1 and 2, $\lambda_1$, which has the smallest available label index, is assigned to optical path 4. Fifth, consider wavelength assignment for optical path 5. As optical 5 shares optical fiber with optical paths 2, 3, and 4, $\lambda_2$, $\lambda_3$, and $\lambda_1$, are not available for optical path 5. Therefore, $\lambda_4$ is assigned to optical path 5. As a result, the required number of wavelengths is 4, as shown in Figure 6.3.

Figure 6.4 shows another example, where wavelengths are assigned sequentially in decreasing order of the number of other optical paths with which the optical path shares the same optical fiber(s). Optical path 3 shares optical fibers with four optical paths, optical paths 2 and 5 share optical fibers with three optical paths, and optical paths 1 and 4 share optical fibers with two optical paths. Therefore, wavelengths are assigned sequentially in the order of paths 3, 2, 5, 1, and 4. First, $\lambda_1$ is assigned to optical path 3. Second, consider wavelength assignment for optical path 2. As optical path 2 shares optical fiber as optical path 3, $\lambda_2$ is assigned to optical path 2. Third, consider wavelength assignment for optical path 5. As optical path 5 shares optical fiber with optical paths 2 and 3, $\lambda_1$ is assigned to optical path 5. Fourth, consider wavelength assignment for optical path 1. As optical path 1 shares optical fiber with optical paths 2 and 3, but does not share any optical fiber with optical path 5, $\lambda_3$ is assigned to optical path 5. Fifth, consider to assign a wavelength to optical path 4. As optical path 4 shares optical fiber with optical paths 3 and 5, but not with optical paths 1 and 2, $\lambda_2$ is assigned to optical path 4. As a result, the required number of wavelengths is 3, as shown in Figure 6.4.

Thus, the required number of wavelengths depends on the order of paths to which wavelengths are assigned. It is necessary to solve an optimization problem to minimize the required number of wavelengths to be assigned in an optical path network.

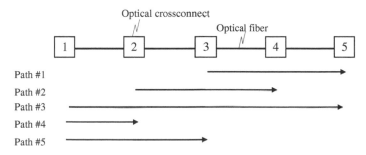

Figure 6.2: Example of optical path requests.

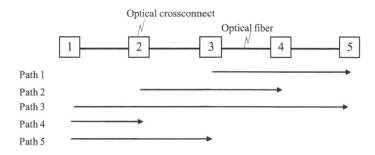

Figure 6.3: Wavelength assignment of optical paths (example 1).

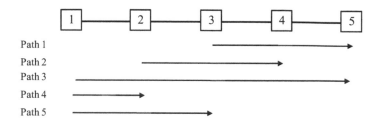

Figure 6.4: Wavelength assignment of optical paths (example 2).

## 6.2   Graph coloring problem

The wavelength assignment problem can be dealt with as a graph coloring problem [1–3]. In the wavelength assignment problem, optical path requests and their routes are given. Each requested optical path corresponds to a vertex in the graph in the graph coloring problem. If two optical paths share an optical fiber, an edge is established between the two corresponding nodes. Otherwise, no edge is established between them. When two vertices are connected by an edge, they are adjacent. A wavelength corresponds to a color in the graph. The graph coloring problem assigns a color to each node while satisfying the constraint that the same color is not assigned to adjacent vertices.

An algorithm to construct the above graph is as follows:

Step 1: Initialize
Initialize the set of vertices $V$ and the set of edges $E$.
$V \leftarrow \{\emptyset\}$, $E \leftarrow \{\emptyset\}$.

Step 2: Vertex generation
Generate vertex $v$, which corresponds to each optical path, and then add $v$ to $V$. This procedure is applied to all optical paths.

Step 3: Edge establishment—
Establish edge $(v, w)$ between $v \in V$ and $w \in V$ if the two optical paths corresponding to vertices $v$ and $w$ pass through the same optical fiber.

For the optical path requested in Figure 6.2, the corresponding graph for the graph coloring problem is shown in Figure 6.5. Vertex $v_i$ corresponds to optical path $i$. If optical paths $i$ and $j$ share the same optical fiber, edge $(v_i, v_j)$ is established between $v_i$ and $v_j$. If $(v_i, v_j)$ exists, the same color cannot be assigned to both vertices.

## 6.3   Integer linear programming

To formulate the graph coloring problem as an ILP problem, the following terminologies are defined. Let $W$ be a set of $\lambda$, where $W = \{\lambda_1, \lambda_2, \cdots, \lambda_{|W|}\}$.

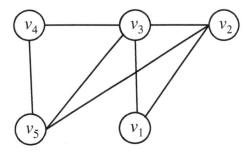

Figure 6.5: Graph construction.

Let $x_v^\lambda$ and $y_\lambda$, be binary variables. If $\lambda$ is assigned to a path corresponding $v$, $x_v^\lambda = 1$, otherwise $x_v^\lambda = 0$. If $\lambda$ is used at least one time, $y_\lambda = 1$; otherwise $y_\lambda = 0$.

The graph coloring problem is formulated as an ILP problem as follows:

$$\text{Objective} \quad \min \sum_{\lambda \in W} y_\lambda \tag{6.1a}$$

$$\text{Constraints} \quad \sum_{\lambda \in W} x_v^\lambda = 1 \quad \forall v \in V \tag{6.1b}$$

$$x_v^\lambda + x_{v'}^\lambda \le y_\lambda \quad \forall (v, v') \in E, \forall \lambda \in W \tag{6.1c}$$

$$y_{\lambda_i} \ge y_{\lambda_{i+1}} (i = 1, 2, \cdots, |W| - 1) \tag{6.1d}$$

$$y_\lambda \in \{0, 1\} \quad \forall \lambda \in W \tag{6.1e}$$

$$x_v^\lambda \in \{0, 1\} \quad \forall v \in V, \quad \forall \lambda \in W \tag{6.1f}$$

Eq. (6.1a) expresses the objective function that minimizes the required number of wavelengths, or colors. Eq. (6.1b) indicates that each optical path uses only one wavelength. Eq. (6.1c) ensures that two adjacent vertices must receive different colors. In other words, this constraint prevents two optical paths sharing the same link from being assigned the same wavelength. In addition, Eq. (6.1c) also indicates that $x_v^\lambda$ must not exceed $y_\lambda$ for all $v \in V$. This means that if $v \in V$ such as $x_v^\lambda = 1$ exists, $y_\lambda$ must be set to 1. Eq. (6.1d) states that wavelengths are used in ascending order of wavelength index $i \in W$. The last two constraints are binary constraints on variable $x_v^\lambda$ and $y_\lambda$.

Eqs. (6.1a)–(6.1f) are separately written in the model file as shown in Listing 6.1 and the input file that represents the wavelength assignment problem in Figure 6.2 as shown in Listing 6.2. Figure 6.5 show a corresponding graph to the wavelength assignment problem.

Listing 6.1: Model file: graph-color-gen.mod

```
1  /* graph-color-gen.mod */
2
3  /* Given parameters */
4  param N integer , >0 ;
```

```
 5
 6   set V := 1..N ;
 7   set E within {V,V} ;
 8   set W := 1..N ;
 9   set W1 := 1..N-1 ;
10
11   param AM{E} ;
12
13   /* Decision variables */
14   var y{W} binary ;
15   var x{V,W} binary ;
16
17   /* Objective function */
18   minimize NUM_COLOR: sum{i in W} y[i] ;
19
20
21   /* Constraints */
22   s.t. X{v in V}:
23        sum{i in W} x[v,i]= 1 ;
24   s.t. XX{v1 in V, v2 in V, i in W: AM[v1,v2]=1}:
25        x[v1,i]+x[v2,i] <= y[i] ;
26   s.t. YY{i in W1}:
27        y[i] >= y [i+1] ;
28   end ;
```

In Listing 6.1, line 3 defines the type of parameter $N$, which indicates the number of possible wavelengths, $|W|$. Line 11 defines the type of parameter AM, which indicates an adjacency matrix. Its element is set to 1 if an edge exists in the graph, otherwise 0.

Listing 6.2: Input file: graph-color-gen1.dat

```
 1   /* graph-color-gen1.dat */
 2
 3   param N := 5 ;
 4
 5   param : E : AM :=
 6   1 1 0
 7   1 2 1
 8   1 3 1
 9   1 4 0
10   1 5 0
11   2 1 1
12   2 2 0
13   2 3 1
14   2 4 0
15   2 5 1
16   3 1 1
17   3 2 1
18   3 3 0
19   3 4 1
20   3 5 1
21   4 1 0
22   4 2 0
23   4 3 1
24   4 4 0
25   4 5 1
26   5 1 0
27   5 2 1
28   5 3 1
29   5 4 1
30   5 5 0
31   ;
32   end;
```

Lines 3–30 of the input file in Listing 6.2 define the values of parameters $N$ and $AM$. Line 3 sets $N = 5$. Lines 5–30 set the adjacency matrix representing Figure 6.5.

After the program is run using 'glpsol', $v_1$, $v_2$, $v_3$, $v_4$, and $v_5$ receive $\lambda_1$, $\lambda_2$, $\lambda_3$, $\lambda_2$, and $\lambda_1$, respectively. The minimum value, which is the required number of wavelength, is 3.

## 6.4   Largest degree first

When the network size becomes large, the complexity of the ILP computations presented in Section 6.3 increases and it becomes difficult to solve it in a practical time. Section 6.4 introduces a heuristic algorithm to solve the problem. It is called the largest degree first (LDF) algorithm.

LDF [4] attempts to color in descending order of degree, where degree is the number of edges connected to the node. LDF works in the following way:

Step 1: Select the uncolored vertex with largest degree.

Step 2: Choose the minimum indexed color from the colors that are not used by the adjacent vertices.

Step 3: Color the selected vertex using the color chosen in Step 2.

Step 4: If all vertices are colored, LDF stops. Otherwise, LDF returns to Step 1.

LDF is a sequential coloring heuristic that attempts to color vertices on the basis of specified order by using the minimum indexed color that is not used by the adjacent vertices. In sequential coloring, if a vertex receives a particular color once, its color remains unchanged thereafter.

Figure 6.6 explains how LDF assigns colors to each vertex, as shown in Figure 6.5, step by step. In the LDF process, vertices are ordered as $v_3$, $v_2$, $v_5$, $v_1$, $v_4$, in descending order of degree. Note that, in case of tie-breaking on the degree, we order them randomly. First, LDF first selects $v_3$ with the largest degree. LDF colors $v_3$ using the minimum indexed wavelength $\lambda_1$ among unused wavelengths, as shown in Figure 6.6(a). Second, LDF selects $v_2$ with the second largest degree. $\lambda_1$ is not available because adjacent vertex $v_3$ uses $\lambda_1$. Therefore, LDF chooses $\lambda_2$, which is the minimum indexed color from the colors that are not used by the adjacent vertices, as shown in Figure 6.6(b). Third, LDF selects $v_5$ with the third largest degree. $\lambda_1$ and $\lambda_2$ are not available because adjacent vertices $v_3$ and $v_2$ use $\lambda_1$ and $\lambda_2$, respectively. LDF chooses $\lambda_3$, as shown in Figure 6.6(c). Fourth, LDF selects $v_1$ with the fourth largest degree. $\lambda_1$ and $\lambda_2$ are not available because adjacent vertices $v_3$ and $v_2$ use $\lambda_1$ and $\lambda_2$, respectively. LDF chooses $\lambda_3$, as shown in Figure 6.6(d). Fifth, LDF selects $v_4$. $\lambda_1$ and $\lambda_3$ are not available because adjacent vertices $v_3$ and $v_5$ use $\lambda_1$ and $\lambda_3$, respectively. LDF chooses $\lambda_2$, as shown in Figure 6.6(e).

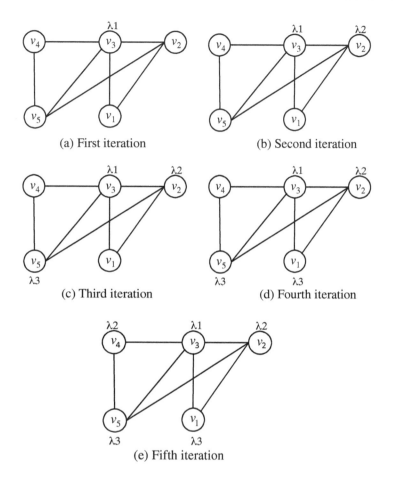

Figure 6.6: Wavelength assignment process of LDF.

After assigning colors to all vertices, the number of colors, or wavelengths, is 3.

## Exercise 6.1

Assign wavelengths for five optical path requests in the network in Figure 6.7.

# Bibliography

[1] H. Zang, J. Jue, and B. Mukherjee, "A review of routing and wavelength assignment approaches for wavelength-routed optical WDM networks," *Optical Networks Magazine,* vol. 1 no. 1 pp. 47–60, 2000.

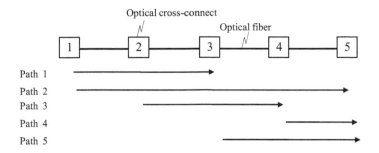

Figure 6.7: Optical path network for wavelength assignment problem.

[2] D. Banerjee and B. Mukherjee, "A practical approach for routing and wavelength assignment in large wavelength-routed optical networks," *IEEE J. Sel. Areas Commun.*, vol. 14, no. 5 pp. 903–908, June 1996.

[3] B. Mukherjee, *Optical WDM Networks*, Springer, New York, 2006.

[4] J. Gross and J. Yellen, *Graph Theory and Its Applications*, CRC Press, Boca Raton, FL, 2006.

# Chapter 7

# Routing and traffic-demand model

Adopting suitable routing can increase the network resource utilization rate and network throughput. Because traffic resources are assigned efficiently, additional traffic can be supported. It also suppresses network congestion and increases robustness in the face of traffic demand fluctuations, most of which are difficult to predict. A traffic demand is defined as the traffic volume that a source node requests to send a destination node. One useful approach to enhancing routing performance is to minimize the maximum link utilization rate, also called the network congestion ratio, of all network links. Minimizing the network congestion ratio leads to an increase in admissible traffic.

This chapter deals with routing problems to minimize the network congestion ratio for several traffic demand models. In general, the more accurately the traffic demand is known, the smaller the network congestion ratio becomes.

## 7.1 Network model

The network is represented as directed graph $G(V, E)$, where $V$ is the set of vertices (nodes) and $E$ is the set of links. Let $Q \subseteq V$ be the set of edges nodes through which traffic is admitted into the network. A link from node $i \in V$ to node $j \in V$ is denoted as $(i, j) \in E$. $c_{ij}$ is the capacity of $(i, j) \in E$. $y_{ij}$ is the link load of $(i, j) \in E$. The traffic demand from node $p$ to node $q$ is denoted as $t_{pq}$. The traffic matrix is denoted as $\boldsymbol{T} = \{t_{pq}\}$. $x_{ij}^{pq}$, where $0 \leq x_{ij}^{pq} \leq 1$ is the portion of the traffic from node $p \in Q$ to node $q \in Q$ routed through $(i, j) \in E$. $\boldsymbol{X}$ is represented as a four-dimensional routing matrix whose element is $x_{ij}^{pq}$, or $\boldsymbol{X} = \{x_{ij}^{pq}\}$, where $0 \leq x_{ij}^{pq} \leq 1$. $\{\boldsymbol{X}\}$ is a set of $\boldsymbol{X}$. $x_{ij}^{pq} > 0$ means $(i, j)$ is a link on one of the routes for $t_{pq}$. $x_{ij}^{pq} = 0$ means that $(i, j)$ is not a link on any route for $t_{pq}$. In the case of $0 < x_{ij}^{pq} < 1$, $t_{pq}$ is split over multiple routes.

The network congestion ratio $r$ refers to the maximum value of all link utilization rates in the network [1]; $r$ is defined by

$$r = \max_{(i,j)\in E} \left\{ \frac{y_{ij}}{c_{ij}} \right\}, \tag{7.1}$$

where $0 \le r \le 1$. The traffic volume of $\frac{1-r}{r}t_{pq}$ is the greatest that can be added to the existing traffic volume of $t_{pq}$ for any pair of source node $p$ and destination node $q$ so that the traffic volume passing though any $(i,j)$ does not exceed $c_{ij}$ under the condition that routing is not changed. After $\frac{1-r}{r}t_{pq}$ is added to $t_{pq}$, the total traffic volume becomes $\frac{1}{r}t_{pq}$ and the updated network congestion becomes 1, which is the upper limit. Maximizing the additional traffic volume of $\frac{1-r}{r}t_{pq}$ is equivalent to minimizing $r$. Thus, minimizing $r$ with routing control is the target in this chapter.

## 7.2    Pipe model

The traffic model that is specified by the exact traffic matrix, $\boldsymbol{T} = \{t_{pq}\}$, is called the *pipe model* [2–4].

An optimal routing formulation with the pipe model to minimize the network congestion ratio was presented in [5]. The network congestion ratio is obtained by solving an LP problem as follows:

$$\min r \tag{7.2a}$$

$$s.t. \quad \sum_{j:(i,j)\in E} x_{ij}^{pq} - \sum_{j:(j,i)\in E} x_{ji}^{pq} = 1,$$

$$\forall p, q \in Q, \text{if } i = p \tag{7.2b}$$

$$\sum_{j:(i,j)\in E} x_{ij}^{pq} - \sum_{j:(j,i)\in E} x_{ji}^{pq} = 0,$$

$$\forall p, q \in Q, i(\neq p, q) \in V \tag{7.2c}$$

$$\sum_{p,q\in Q} t_{pq} x_{ij}^{pq} \le c_{ij} \cdot r, \quad \forall(i,j) \in E \tag{7.2d}$$

$$0 \le x_{ij}^{pq} \le 1, \quad \forall p, q \in Q, (i,j) \in E \tag{7.2e}$$

$$0 \le r \le 1. \tag{7.2f}$$

The decision variables are $r$ and $x_{ij}^{pq}$, and the given parameters are $t_{pq}$ and $c_{ij}$. The objective function in Eq. (7.2a) minimizes the network congestion ratio. Eqs. (7.2b) and (7.2c) are constraints for flow conservation. Eq. (7.2b) states that the total traffic flow ratio outgoing from node $i(= p)$, which is a source node, is 1. Eq. (7.2c) states that the traffic flow incoming to node $i$ must be the traffic outgoing from node $i$ if node $i$ is neither a source nor destination node for the flow. As flow conservation is satisfied at a destination if Eqs. (7.2b) and (7.2c) are satisfied, the condition does not need to be included as a constraint. Eq. (7.2d) indicates that the sum of the fractions of traffic demands transmitted over $(i, j)$ is equal to or less than the network congestion ratio times the total capacity $c_{ij}$ for all links.

The pipe model imposes the condition that $t_{pq}$ is exactly known. However, it is difficult for network operators to know the actual traffic matrix when the network size is large [6–9]. For example, in Internet Protocol (IP) networks, to measure $t_{pq}$ at source node $p$, the node checks all the destination addresses in each IP packet header and counts the number of packets destined for node $q$. It makes the processing load of node $p$ increase.

## 7.3   Hose model

It is easy for network operators to specify the traffic as just the total outgoing/incoming traffic from/to node $p$ and node $q$, The total outgoing traffic from node $p$ is represented as $\sum_q t_{pq} \leq \alpha_p$, where $\alpha_p$ is the maximum rate of traffic that node $p$ can send into the network. The total incoming traffic to node $q$ is represented as $\sum_p t_{pq} \leq \beta_q$, where $\beta_q$ is the maximum rate of traffic that node $q$ can receive from the network. The traffic model that is bounded by $\alpha_p$ and $\beta_q$ is called the *hose model* [2–4]. In the hose model, the traffic demand between each source-destination pair does not need to be specified. Therefore, it is beneficial for network operators to specify a set of traffic conditions in the hose model, especially when the network is large.

The optimal routing problem with the hose model is also formulated by Eq. (7.2a)–(7.2f). The decision variables are $r$ and $x_{ij}^{pq}$, and the given parameters are $t_{pq}$ and $c_{ij}$. However, $t_{pq}$ is bounded by the hose model as follows:

$$\sum_{q \in Q} t_{pq} \leq \alpha_p, \quad p \in Q \tag{7.3a}$$

$$\sum_{p \in Q} t_{pq} \leq \beta_q, \quad q \in Q. \tag{7.3b}$$

Let a set of $\boldsymbol{T}$s that satisfies the conditions specified by a certain traffic demand model be $\{\boldsymbol{T}\}$. In the hose model, $\{\boldsymbol{T}\}$ is specified by Eqs. (7.3a) and (7.3b). Assuming that $\boldsymbol{T}$ is given, the network congestion ratio, which refers to the maximum value of all link utilization rates in the network, is denoted as $r$. Minimizing $r$ with routing control means that admissible traffic is maximized. From the network operator's point of view, routing $\boldsymbol{X}$ should not be changed for stable network operation, even when $\boldsymbol{T}$ is varied within the range of $\{\boldsymbol{T}\}$.

We would like to find the optimal routing that minimizes $r$ for $\boldsymbol{X} \in \{\boldsymbol{X}\}$, and maximizes the minimal $r$ in terms of $\boldsymbol{T} \in \{\boldsymbol{T}\}$. This routing is called *oblivious routing* [8–13].

$$\max_{\boldsymbol{T} \in \{\boldsymbol{T}\}} \min_{\boldsymbol{X} \in \{\boldsymbol{X}\}} r \tag{7.4}$$

To find the optimal $r$ and $\boldsymbol{X}$, first an optimal routing problem to obtain $\min_{\boldsymbol{X} \in \{\boldsymbol{X}\}} r$ is considered, under the condition that $\boldsymbol{T} \in \{\boldsymbol{T}\}$ is given. Then, "$\max_{\boldsymbol{T} \in \{\boldsymbol{T}\}}$" in Eq. (7.4) is incorporated into the problem. For a given $\boldsymbol{T} \in \{\boldsymbol{T}\}$, an optimal routing formulation with the hose model is also presented in Eqs. (7.2a)–(7.2f).

Although Eqs. (7.2a)–(7.2f) can be expressed as an LP problem, they cannot be easily solved as a regular LP problem. Constraint (7.2d) lists every valid combination in $\boldsymbol{T} = \{t_{pq}\}$ bounded by Eqs. (7.3a) and (7.3b). It is impossible to repeatedly solve the LP problems for all possible sets of $\boldsymbol{T} \in \{\boldsymbol{T}\}$. This is because $t_{pq}$ takes a real value, and the number of valid combinations of $t_{pq}$ is infinite. This problem is solved using Chu's property, Property 1 [6, 7].

**Property 1.** $x_{ij}^{pq}$ achieves congestion ratio $\leq r$ for all traffic matrices in $\boldsymbol{T} = \{t_{pq}\}$ constrained by Eqs. (7.3a) and (7.3b) if and only if there exist non-negative parameters $\pi_{ij}(p)$ and $\lambda_{ij}(p)$ for every $(i, j) \in E$ such that
(i) $\sum_{p \in Q} \alpha_p \pi_{ij}(p) + \sum_{p \in Q} \beta_p \lambda_{ij}(p) \leq c_{ij} \cdot r$
for each $(i, j) \in E$
(ii) $x_{ij}^{pq} \leq \pi_{ij}(p) + \lambda_{ij}(q)$ for each $(i, j) \in E$ and every $p, q \in Q$
$\pi_{ij}(p)$ and $\lambda_{ij}(p)$ are produced by the dual theorem when a primal problem is transformed into the dual problem, as described in the following proof [6, 7]. The left term of condition (i) is derived as the objective function of the dual problem, where the objective function of the primal problem is $\sum_{p,q \in Q} x_{ij}^{pq} t_{pq}$. Condition (ii) is obtained as a constraint of the dual problem.
**Proof:**

("only if" direction): Let routing $x_{ij}^{pq}$ have congestion ratio $\leq r$ for all traffic matrices constrained by the hose model. (i.e., $\sum_{p,q \in Q} x_{ij}^{pq} t_{pq} \leq c_{ij} \cdot r$ for all $(i, j)$). The problem of finding $\boldsymbol{T} = \{t_{pq}\}$ that maximizes link load on $(i, j)$ is formulated as the following LP problem:

$$\max \sum_{p,q \in Q} x_{ij}^{pq} t_{pq} \tag{7.5a}$$

$$\text{s.t.} \sum_{q \in Q} t_{pq} \leq \alpha_p, \quad \forall p \in Q \tag{7.5b}$$

$$\sum_{p \in Q} t_{pq} \leq \beta_q, \quad \forall q \in Q \tag{7.5c}$$

The decision variables are $t_{pq}$. The given parameters are $x_{ij}^{pq}$, $\alpha_p$, and $\beta_q$. The dual of the LP problem in Eqs. (7.5a)–(7.5c) for $(i, j)$ is

$$\min \sum_{p \in Q} \alpha_p \pi_{ij}(p) + \sum_{p \in Q} \beta_p \lambda_{ij}(p) \tag{7.6a}$$

$$\text{s.t.} \quad x_{ij}^{pq} \leq \pi_{ij}(p) + \lambda_{ij}(q), \quad \forall p, q \in Q, (i, j) \in E \tag{7.6b}$$

$$\pi_{ij}(p), \lambda_{ij}(p) \geq 0, \quad \forall p, q \in Q, (i, j) \in E \tag{7.6c}$$

The derivation of Eqs. (7.6a)–(7.6c) is described in Appendix A. Because of $\sum_{pq} x_{ij}^{pq} t_{pq} \leq c_{ij} \cdot r$ in Eq. (7.5a), the dual, $\sum_{p \in Q} \alpha_p \pi_{ij}(p) + \sum_{p \in Q} \beta_p \lambda_{ij}(p)$ in Eq. (7.6a), for any $(i, j)$, must have the same optimal value. The optimal value in Eq. (7.6a) should be $\leq c_{ij} \cdot r$. Therefore, the objective function of the dual satisfies (i). Requirement (ii) is satisfied by dual problem constraint (7.6b). ("if" direction): Let $x_{ij}^{pq}$ be a routing and $\boldsymbol{T} = \{t_{pq}\}$ be any valid traffic matrix. Let $\pi_{ij}(p)$ and $\lambda_{ij}(p)$ be the parameters satisfying requirements (i) and (ii). Consider $(i, j) \in E$. From (ii) we have

$$x_{ij}^{pq} \leq \pi_{ij}(p) + \lambda_{ij}(q).$$

Summing over all edge node pairs $(p, q)$, we have

$$
\begin{aligned}
\sum_{p,q \in Q} x_{ij}^{pq} t_{pq} &\leq \sum_{p,q \in Q} [\pi_{ij}(p) + \lambda_{ij}(q)] t_{pq} \\
&= \sum_{p \in Q} \pi_{ij}(p) \sum_{q \in Q} t_{pq} + \sum_{q \in Q} \lambda_{ij}(q) \sum_{p \in Q} t_{pq} \\
&\leq \sum_{p \in Q} \pi_{ij}(p) \alpha_p + \sum_{p \in Q} \lambda_{ij}(p) \beta_p.
\end{aligned}
$$

$$(7.7)$$

The last equality is obtained using the constraints of the hose model. From (i), we have

$$
\sum_{p,q \in Q} x_{ij}^{pq} t_{pq} \leq \sum_{p \in Q} \pi_{ij}(p) \alpha_p + \sum_{p \in Q} \lambda_{ij}(p) \beta_p
$$

$$(7.8)$$

$$\leq c_{ij} \cdot r.$$

This indicates that for any traffic matrix constrained by the hose model, the load on any link is at most $r$. ∎

Property 1 allows us to replace constraint (7.2d) in Eqs. (7.2a)–(7.2f), where $t_{pq}$ is bounded by Eqs. (7.3a) and (7.3b), with requirements (i) and (ii) in Property 1; this transforms the formulation as follows:

$$\min r \tag{7.9a}$$

$$s.t. \quad \sum_{j:(i,j)\in E} x_{ij}^{pq} - \sum_{j:(j,i)\in E} x_{ji}^{pq} = 1,$$

$$\forall p, q \in Q, \text{if } i = p \tag{7.9b}$$

$$\sum_{j:(i,j)\in E} x_{ij}^{pq} - \sum_{j:(j,i)\in E} x_{ji}^{pq} = 0,$$

$$\forall p, q \in Q, i(\neq p, q) \in V \tag{7.9c}$$

$$\sum_{p\in Q} \alpha_p \pi_{ij}(p) + \sum_{p\in Q} \beta_p \lambda_{ij}(p) \leq c_{ij} \cdot r, \qquad \forall (i,j) \in E \tag{7.9d}$$

$$x_{ij}^{pq} \leq \pi_{ij}(p) + \lambda_{ij}(q), \qquad \forall p, q \in Q, (i,j) \in E \tag{7.9e}$$

$$\pi_{ij}(p), \lambda_{ij}(p) \geq 0, \qquad \forall p, q \in Q, (i,j) \in E \tag{7.9f}$$

$$0 \leq x_{ij}^{pq} \leq 1, \quad \forall p, q \in Q, (i,j) \in E \tag{7.9g}$$

$$0 \leq r \leq 1. \tag{7.9h}$$

The decision variables are $r$, $x_{ij}^{pq}$, $\pi_{ij}(p)$, and $\lambda_{ij}(p)$, and the given parameters are $c_{ij}$, $\alpha_p$, and $\beta_q$. $\pi_{ij}(p)$ is the ratio of traffic on $(i,j)$ outgoing from node $p$, and $\lambda_{ij}(q)$ is the ratio of traffic on $(i,j)$ incoming to node $q$ [14]. Eq. (7.2d) and Eqs. (7.3a) and (7.3b) are replaced by Eqs. (7.9d)–(7.9f). By introducing the variables of $\pi_{ij}(p)$ and $\lambda_{ij}(p)$, Eqs. (7.3a)-(7.3b) are incorporated in this optimization problem. Eqs. (7.9a)-(7.9h) represent a regular LP problem and can be solved optimally with a standard LP solver.

The hose model is more beneficial than the pipe model because the hose model is easier for network operators to specify. However, the worst-case network congestion ratio of the hose model is larger than that of the pipe model, as all the possible traffic demands bounded by the hose model must be considered in routing selection. The degree of difference between both network congestion ratios, using five sample networks as shown in Figure 7.1, is examined. We obtained the average values of the normalized network congestion ratios for 100 randomly generated conditions of link capacities and traffic demands for the sample networks. $\alpha_p = \sum_q t_{pq}$ and $\beta_q = \sum_p t_{pq}$ are set. Figure 7.2 compares the network congestion ratios of the pipe and hose models, which are normalized by that of the hose model. It is observed that the pipe model offers a 30 to 40% lower network congestion ratio than the hose model for the sample networks, but at the cost of requiring an accurate traffic matrix.

## 7.4   HSDT model

Sections 7.4 and 7.5 describe traffic demand models that narrow the range to traffic conditions specified by the hose model so that the worst-case network

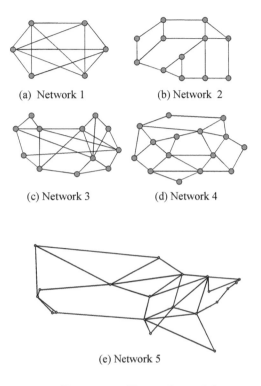

(a) Network 1(b) Network 2

(c) Network 3(d) Network 4

(e) Network 5

Figure 7.1: Network models.

congestion ratio can be reduced, while the advantage of the hose model is preserved.

The hose model has a weakness in that its routing performance is much lower than that of the pipe model. This is because all possible sets of $T = \{t_{pq}\}$ must be considered in the hose model, while the exact $T = \{t_{pq}\}$ is known in the pipe model. Therefore, it is desirable for network operators to narrow the range of traffic conditions specified by the hose model so as to enhance its routing performance.

Network operators are able to impose additional bounds on the hose model from their operational experience and past traffic data as described below. First, the range of errors of the traffic matrix can be estimated. It is reported that traffic estimations based on link traffic measurements are in error by 20% or more [15]. There are several studies on estimating the traffic matrix [16–18]. To estimate the traffic matrix, network operators do not need to measure each traffic demand $d_{pq}$. Instead, they only have to measure the traffic loads on each link in the network. Using the measured link loads, the traffic matrix can be estimated with certain accuracy, but it is inevitable that the estimated traffic matrix includes some errors. Second, traffic fluctuations from the network operational history can be predicted. Third, the network operators know the

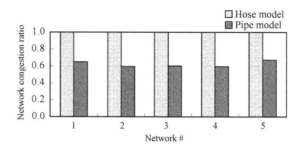

Figure 7.2: Comparisons of congestion ratio between pipe and hose models for sample networks. (©2009 IEEE, Ref. [19].)

trends of traffic demands. For example, traffic demands between large cities are much larger than those of small cities. Thus, the additional bounds are expressed by $\delta_{pq} \leq t_{pq} \leq \gamma_{pq}$ for each pair of source node $p$ and destination $q$. $\gamma_{pq}$ and $\delta_{pq}$ are the upper and lower bounds of $t_{pq}$, respectively [19, 20]. We call this model, the hose model with additional bounds, the *Hose model with bounds of Source-Destination Traffic demands* (HSDT model).

Thus, $t_{pq}$ is bounded by the HSDT model as follows.

$$\sum_{q \in Q} t_{pq} \leq \alpha_p, \quad p \in Q \tag{7.10a}$$

$$\sum_{p \in Q} t_{pq} \leq \beta_q, \quad q \in Q \tag{7.10b}$$

$$\delta_{pq} \leq t_{pq} \leq \gamma_{pq}, \quad p, q \in Q \tag{7.10c}$$

The HSDT model offers better routing performance than the hose model, by narrowing the range of traffic conditions specified by the hose model. The features of the pipe, hose, and HSDT models are schematically shown in Figure 7.3. When the range of $t_{pq}$ is not exactly determined, a large range of $t_{pq}$ is set, that is, a conservative approach, to avoid any congestion. This may be still effective, because $\alpha_p$ and $\beta_q$ also bound the traffic demands.

To find the optimal routing that minimizes the worst-case network congestion ratio in the HSDT model, we solve the optimal routing problem presented in Eq. 7.4, where $\{T\}$ is specified by Eqs. (7.10a)–(7.10c). In the same way as the approach for the hose model, the optimization problem in Eq. 7.4 is transformed into the LP problem that does not include $t_{pq}$, using the following property, Property 2.

**Property 2.** $x_{ij}^{pq}$ achieves congestion ratio $\leq r$ for all traffic matrices in $T = \{t_{pq}\}$ constrained by Eqs. (7.10a)-(7.10c) if and only if there exist non-

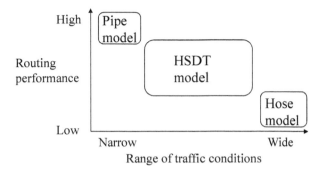

Figure 7.3: Features of pipe, hose, and HSDT models.

negative parameters $\pi_{ij}(p)$, $\lambda_{ij}(p)$, $\eta_{ij}(p,q)$, and $\theta_{ij}(p,q)$ for every $(i,j) \in E$ such that

(i) $\sum_{p \in Q} \alpha_p \pi_{ij}(p) + \sum_{p \in Q} \beta_p \lambda_{ij}(p)$
$+ \sum_{p \in Q} \sum_{q \in Q} [\gamma_{pq} \eta_{ij}(p,q) - \delta_{pq} \theta_{ij}(p,q)] \leq c_{ij} \cdot r$ for each $(i,j) \in E$,
(ii) $x_{ij}^{pq} \leq \pi_{ij}(p) + \lambda_{ij}(q) + \eta_{ij}(p,q) - \theta_{ij}(p,q)$ for each $(i,j) \in E$ and every $p, q \in Q$.

$\pi_{ij}(p)$, $\lambda_{ij}(p)$, $\eta_{ij}(p,q)$, and $\theta_{ij}(p,q)$ are produced by the dual theorem when a primal problem is transformed into the dual problem. The left term of condition (i) is derived as the objective function of the dual problem, where the objective function of the primal problem is $\sum_{p,q \in Q} x_{ij}^{pq} t_{pq}$. Condition (ii) is obtained as a constraint of the dual problem.

Property 2 is proved in the same way as the proof of Property 1 by considering the following primal and dual problems (see Exercise 7.1). The problem of finding $T = \{t_{pq}\}$ that maximizes link load on $(i,j)$ is formulated as the following LP problem, which is the primal problem:

$$\max \sum_{p,q \in Q} x_{ij}^{pq} t_{pq} \tag{7.11a}$$

$$\text{s.t.} \sum_{q \in Q} t_{pq} \leq \alpha_p, \quad \forall p \in Q \tag{7.11b}$$

$$\sum_{p \in Q} t_{pq} \leq \beta_q, \quad \forall q \in Q \tag{7.11c}$$

$$\delta_{pq} \leq t_{pq} \leq \gamma_{pq}, \quad \forall p, q \in Q. \tag{7.11d}$$

The decision variables are $t_{pq}$. The given parameters are $x_{ij}^{pq}$, $\alpha_p$, $\beta_q$, $\delta_{pq}$, and

$\gamma_{pq}$. The dual of the LP problem in Eqs. (7.11a)–(7.11d) for $(i,j)$ is

$$\min \sum_{p \in Q} \alpha_p \pi_{ij}(p) + \sum_{p \in Q} \beta_p \lambda_{ij}(p)$$

$$+ \sum_{p,q \in Q} [\gamma_{pq} \eta_{ij}(p,q) - \delta_{pq} \theta_{ij}(p,q)], \tag{7.12a}$$

$$s.t. \quad x_{ij}^{pq} \leq \pi_{ij}(p) + \lambda_{ij}(q) + \eta_{ij}(p,q) - \theta_{ij}(p,q),$$

$$\forall p,q \in Q, (i,j) \in E \tag{7.12b}$$

$$\pi_{ij}(p), \lambda_{ij}(p), \eta_{ij}(p,q), \theta_{ij}(p,q) \geq 0,$$

$$\forall p,q \in Q, (i,j) \in E. \tag{7.12c}$$

The derivation of Eqs. (7.12a)–(7.12c) is described in Appendix B. Because of $\sum_{pq} x_{ij}^{pq} t_{pq} \leq c_{ij} \cdot r$ in Eq. (7.11a), the dual, $\sum_{p \in Q} \alpha_p \pi_{ij}(p) + \sum_{p \in Q} \beta_p \lambda_{ij}(p) + \sum_{p,q \in Q} [\gamma_{pq} \eta_{ij}(p,q) - \delta_{pq} \theta_{ij}(p,q)$ in Eq. (7.12a), for any $(i,j)$, must have the same optimal value.

Property 2 allows us to replace constraint (7.2d) in Eqs. (7.2a)–(7.2f), where $t_{pq}$ is bounded by Eqs. (7.10a)–(7.10c), with requirements (i) and (ii) in Property 2; this transforms the formulation as follows:

$$\min r \tag{7.13a}$$

$$s.t. \quad \sum_{j:(i,j) \in E} x_{ij}^{pq} - \sum_{j:(j,i) \in E} x_{ji}^{pq} = 1,$$

$$\forall p,q \in Q, \text{if } i = p \tag{7.13b}$$

$$\sum_{j:(i,j) \in E} x_{ij}^{pq} - \sum_{j:(j,i) \in E} x_{ji}^{pq} = 0,$$

$$\forall p,q \in Q, i(\neq p,q) \tag{7.13c}$$

$$\sum_{p \in Q} \alpha_p \pi_{ij}(p) + \sum_{p \in Q} \beta_p \lambda_{ij}(p)$$

$$+ \sum_{p,q \in Q} [\gamma_{pq} \eta_{ij}(p,q) - \delta_{pq} \theta_{ij}(p,q)] \leq c_{ij} \cdot r,$$

$$\forall (i,j) \in E \tag{7.13d}$$

$$x_{ij}^{pq} \leq \pi_{ij}(p) + \lambda_{ij}(q) + \eta_{ij}(p,q) - \theta_{ij}(p,q),$$

$$\forall p,q \in Q, (i,j) \in E \tag{7.13e}$$

$$\pi_{ij}(p), \lambda_{ij}(p), \eta_{ij}(p,q), \theta_{ij}(p,q) \geq 0,$$

$$\forall p,q \in Q, (i,j) \in E \tag{7.13f}$$

$$0 \leq x_{ij}^{pq} \leq 1, \quad \forall p,q \in Q, (i,j) \in E \tag{7.13g}$$

$$0 \leq r \leq 1. \tag{7.13h}$$

The decision variables are $r$, $x_{ij}^{pq}$, $\pi_{ij}(p)$, $\lambda_{ij}(p)$, $\eta_{ij}(p,q)$, and $\theta_{ij}(p,q)$, and the given parameters are $c_{ij}$, $\alpha_p$, $\beta_q$, $\delta_{pq}$, and $\gamma_{pq}$. Eq. (7.2d) and Eqs. (7.10a)–(7.10c) are replaced by Eqs. (7.13d)–(7.13f). By introducing the variables of

$\pi_{ij}(p)$, $\lambda_{ij}(p)$, $\eta_{ij}(p,q)$, and $\theta_{ij}(p,q)$, Eqs. (7.10a)–(7.10c) are incorporated in this optimization problem. Eqs. (7.13a)–(7.13h) represent a regular LP problem and can be solved optimally with a standard LP solver.

The network congestion ratios, $r$, of the different models, which are the HSDT mode, the pipe model, and the hose model, are compared. The network congestion ratios of the HSDT and pipe models are normalized by that of the hose model. The normalized network congestion ratios for the HSDT, pipe, and hose models are denoted as $r_{HSDT}$, $r_P$, and $r_H (= 1.0)$, respectively. The conditions of traffic demands and link capacities are the same as those presented in Section 7.3. By using randomly generated $t_{pq}$, the following parameters are set as follows: $\alpha_p = \sum_q t_{pq}$, $\beta_q = \sum_p t_{pq}$, $\gamma_{pq} = \frac{1}{\mu} t_{pq}$, and $\delta_{pq} = \nu t_{pq}$. $\mu$ and $\nu$ are parameters to express $\gamma_{pq}$ and $\delta_{pq}$, respectively, where $0 < \mu \leq 1$ and $0 \leq \nu \leq 1$. $(\mu, \nu) \to (1,1)$ and $(\mu, \nu) \to (0,0)$ indicate that the HSDT model approaches pipe and hose model performance, respectively.

Figure 7.4 show comparisons of the network congestion ratios of the different models, for network 3 shown in Figure 7.1(c). It indicates that $r_P \leq r_{HSDT} \leq r_H = 1.0$ is satisfied for all sets of $(\mu, \nu)$, where $(\mu, \nu)$ is defined in the beginning of this section. When $(\mu, \nu)$ is close to $(1,1)$ and $(0,0)$, the network congestion ratio of the HSDT model is close to those of the pipe model and the hose model, respectively. Note that the network congestion ratio of the HSDT model does not depend on $\mu$ when $\mu$ is small enough. This means that the traffic matrix $\boldsymbol{T} = \{t_{pq}\}$ is bounded by only $\alpha_p$ and $\beta_q$, not by $\gamma_{pq}$. If network operators can specify an appropriate set of $(\mu, \nu)$ so that they can narrow the possible range of $\boldsymbol{T} = \{t_{pq}\}$, the network congestion ratio of the HSDT model is reduced compared to that of the hose model by solving our LP problem formulation. This is a key advantage of the HSDT model. For example, when $(\mu, \nu)=(0.8, 0.8)$ is set in Figure 7.4, the network congestion ratio of the HSDT model is 34% less than that of the hose model. $\mu = 0.8$ and $\nu = 0.8$ means that 25% $(1/0.8 - 1)$ upper-bound margin of $t_{pq}$ and 20% $(1 - 0.8)$ lower-bound margin are considered, respectively.

## 7.5 HLT model

In the HSDT model, it may not always be easy for the network operators to specify the additional bounds on source-destination traffic demands. The existing approach uses the operator's experience and past traffic data in determining the bounds. Therefore, it is desirable for network operators to be able to determine appropriate additional bounds in a more objective manner.

We can use the total traffic passing through links as an additional bound on the hose model, instead of the source-destination traffic demands in the HSDT model. The total traffic volume passing through each link can be measured much more easily than source-destination traffic volume [15]. This is because the former requires only the number of packets passing through the link to be measured, while the latter requires the number of packets for each destination

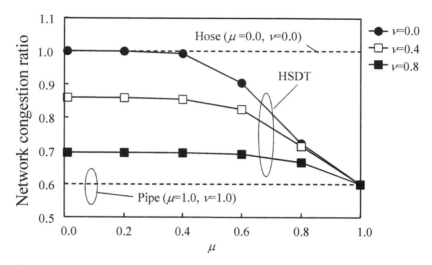

Figure 7.4: Comparison of congestion ratios. (©2009 IEEE, Ref. [19].)

to be measured, which requires that the IP destination address in the IP packet header be extracted and read.

We represent the additional upper bound, the total traffic passing through link $(i, j)$, as $\sum_{p,q \in Q} a_{ij}^{pq} t_{pq} \leq y_{ij}$, where $y_{ij}$ is the maximum rate of traffic accommodated in link $(i, j)$, and $a_{ij}^{pq}$ is the portion of the traffic from node $p$ to node $q$ routed through link $(i, j)$. Network operators are able to know the routing information of $a_{ij}^{pq}$, which is determined by them, and to specify $y_{ij}$ by measuring the total traffic passing through link $(i, j)$. This model, in which bounds derived from the measured total traffic on link $(i, j)$ are added to the hose model, is called *the Hose model with bounds of Link Traffic volume* (HLT model) [21]. The HLT model offers better routing performance than the hose model, by narrowing the range of traffic conditions. In the HLT model, to perform the routing optimization, there are two phases. In the first phase, network operators configure routing in the network as an initial condition. For example, the shortest path routing policy may be adopted, and $a_{ij}^{pq}$ and $y_{ij}$ are determined. In the second phase, using $a_{ij}^{pq}$ and $y_{ij}$ determined by the first phase, routing optimization is performed.

Thus, $t_{pq}$ is bounded by the HSDT model as follows:

$$\sum_{q \in Q} t_{pq} \leq \alpha_p, \quad p \in Q \tag{7.14a}$$

$$\sum_{p \in Q} t_{pq} \leq \beta_q, \quad q \in Q \tag{7.14b}$$

$$\sum_{p,q \in Q} a_{ij}^{pq} t_{pq} \leq y_{ij}, \quad (i, j) \in E \tag{7.14c}$$

To find the optimal routing that minimizes the worst-case network congestion ratio in the HLT model, we solve the optimal routing problem presented
in Eq. (7.4), where $\{T\}$ is specified by Eqs. (7.14a–7.14c). In the same way
as the approach for the hose model and the HSDT model, the optimization
problem in Eq. 7.4 is transformed into the LP problem that does not include
$t_{pq}$, by using the following property, Property 3.

**Property 3.** $x_{ij}^{pq}$ achieves congestion ratio $\leq r$ for all traffic matrices in
$T = \{t_{pq}\}$ constrained by Eqs. (7.14a)–(7.14c) if and only if there exist non-
negative parameters $\pi_{ij}(p)$, $\lambda_{ij}(p)$, and $\theta_{ij}(s,t)$ for every $(i,j) \in E$ such that
(i) $\sum_{p \in Q} \alpha_p \pi_{ij}(p) + \sum_{p \in Q} \beta_p \lambda_{ij}(p)$
$+ \sum_{(s,t) \in E} y_{st} \theta_{ij}(s,t) \leq c_{ij} \cdot r$ for each $(i,j) \in E$
(ii) $x_{ij}^{pq} \leq \pi_{ij}(p) + \sum_{(s,t) \in E} a_{st}^{pq} \theta_{ij}(s,t)$ for each $(i,j) \in E$ and every $p,q \in Q$.

$\pi_{ij}(p)$, $\lambda_{ij}(p)$, and $\theta_{ij}(s,t)$ are produced by the dual theorem when a primal
problem is transformed into the dual problem. The left term of condition (i)
is derived as the objective function of the dual problem, where the objective
function of the primal problem is $\sum_{p,q \in Q} x_{ij}^{pq} t_{pq}$. Condition (ii) is obtained as
a constraint of the dual problem.

Property 3 is proved in the same way of the proof by Property 1 by considering the following primal and dual problems (see Exercise 7.2). The problem
of finding $T = \{t_{pq}\}$ that maximizes link load on $(i,j)$ is formulated as the
following LP problem, which is the primal problem:

$$\max \sum_{p,q \in Q} x_{ij}^{pq} t_{pq} \tag{7.15a}$$

$$\text{s.t.} \sum_{q \in Q} t_{pq} \leq \alpha_p, \quad \forall p \in Q \tag{7.15b}$$

$$\sum_{p \in Q} t_{pq} \leq \beta_q, \quad \forall q \in Q \tag{7.15c}$$

$$\sum_{p,q \in Q} a_{ij}^{pq} t_{pq} \leq y_{ij}, \quad \forall (i,j) \in E. \tag{7.15d}$$

The decision variables are $t_{pq}$. The given parameters are $x_{ij}^{pq}$, $\alpha_p$, $\beta_q$, $a_{ij}^{pq}$, and
$y_{ij}$. The dual of the LP problem in Eqs. (7.15a)–(7.15d) for $(i,j)$ is

$$\min \sum_{p \in Q} \alpha_p \pi_{ij}(p) + \sum_{p \in Q} \beta_p \lambda_{ij}(p) + \sum_{(s,t) \in E} \theta_{ij}(s,t) y_{st} \tag{7.16a}$$

$$\text{s.t.} \ \pi_{ij}(p) + \lambda_{ij}(q) + \sum_{(s,t) \in E} a_{st}^{pq} \theta_{ij}(s,t) \geq x_{ij}^{pq}, \quad \forall p,q \in Q, (i,j) \in E \tag{7.16b}$$

$$\pi_{ij}(p), \lambda_{ij}(p) \geq 0, \quad \forall p \in Q, (i,j) \in E \tag{7.16c}$$

$$\theta_{ij}(s,t) \geq 0, \quad \forall (i,j), (s,t) \in E. \tag{7.16d}$$

The derivation of Eqs. (7.16a)–(7.16d) is described in Appendix C. Because of $\sum_{pq} x_{ij}^{pq} t_{pq} \leq c_{ij} \cdot r$ in Eq. (7.15a), the dual, $\sum_{p \in Q} \alpha_p \pi_{ij}(p) + \sum_{p \in Q} \beta_p \lambda_{ij}(p) + \sum_{(s,t) \in E} \theta_{ij}(s,t) y_{st}$ in Eq. (7.16a), for any $(i,j)$, must have the same optimal value.

Property 3 allows us to replace constraint (7.2d) in Eqs. (7.2a)–(7.2f), where $t_{pq}$ is bounded by Eqs. (7.14a)–(7.14c), with requirements (i) and (ii) in Property 3; this transforms the formulation as follows:

$$\min r \tag{7.17a}$$

$$\text{s.t.} \quad \sum_{j:(i,j) \in E} x_{ij}^{pq} - \sum_{j:(j,i) \in E} x_{ji}^{pq} = 1,$$

$$\forall p, q \in Q, \text{if } i = p \tag{7.17b}$$

$$\sum_{j:(i,j) \in E} x_{ij}^{pq} - \sum_{j:(j,i) \in E} x_{ji}^{pq} = 0,$$

$$\forall p, q \in Q, i (\neq p, q) \tag{7.17c}$$

$$\sum_{p \in Q} \pi_{ij}(p)\alpha_p + \sum_{p \in Q} \lambda_{ij}(p)\beta_p$$

$$+ \sum_{(s,t) \in E} \theta_{ij}(s,t) y_{st} \leq c_{ij} \cdot r, \quad \forall (i,j) \in E \tag{7.17d}$$

$$\pi_{ij}(p) + \lambda_{ij}(p) + \sum_{(s,t) \in E} a_{st}^{pq} \theta_{ij}(s,t) \geq x_{ij}^{pq},$$

$$\forall (i,j) \in E, p, q \in Q \tag{7.17e}$$

$$\pi_{ij}(p), \lambda_{ij}(p) \geq 0, \quad \forall (i,j) \in E, p \in Q \tag{7.17f}$$

$$\theta_{ij}(s,t) \geq 0, \quad \forall (s,t), (i,j) \in E \tag{7.17g}$$

$$0 \leq x_{ij}^{pq} \leq 1, \quad \forall p, q \in Q, (i,j) \in E \tag{7.17h}$$

$$0 \leq r \leq 1. \tag{7.17i}$$

The decision variables are $r$, $x_{ij}^{pq}$, $\pi_{ij}(p)$, $\lambda_{ij}(p)$, and $\theta_{ij}(s,t)$, and the given parameters are $c_{ij}$, $\alpha_p$, $\beta_p$, and $y_{st}$. Eq. (7.2d) and Eqs. (7.14a)–(7.14c) are replaced by Eqs. (7.17d)–(7.17g). By introducing the variables of $\pi_{ij}(p)$, $\lambda_{ij}(p)$, and $\theta_{ij}(s,t)$, Eqs. (7.14a)–(7.14c) are incorporated in this optimization problem. Eqs. (7.17a)–(7.17i) represent a regular LP problem and can be solved optimally with a standard LP solver.

The network congestion ratios of the different models, which are the HLT model, the pipe model, and the hose model, are compared. The network congestion ratios of the HLT and pipe models are normalized by that of the hose model. The conditions of traffic demands and link capacities are the same as those presented in Section 7.3. Using randomly generated $t_{pq}$, $\alpha_p = \sum_q t_{pq}$ and $\beta_q = \sum_p t_{pq}$ are set. The portion of traffic $a_{ij}^{pq}$ is determined by assuming that, before the optimization process, each route between source and destination nodes follows the shortest path.

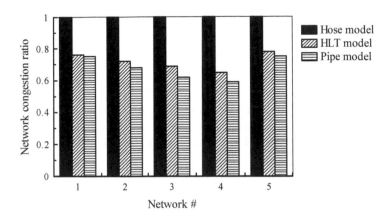

Figure 7.5: Comparison of congestion ratios for pipe, hose and HLT models. (©2011 IEEE, Ref. [21].)

Figure 7.5 shows that the pipe model offers 25 to 40% lower network congestion ratio than the hose model for the sample network, and the HLT model offers 20 to 35% lower network congestion ratio than the hose model. In addition, the differences in the network congestion ratios between the pipe model and the HLT model for the examined sample networks are less than 0.1.

## Exercise 7.1

Consider the network in Figure 7.6. The capacity matrix, $\mathbf{C} = \{c_{ij}\}$, where

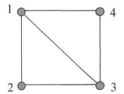

Figure 7.6: Network model.

$c_{ij}$ is the capacity of link $(i, j)$, is given by

$$C = \begin{bmatrix} 0 & 75 & 75 & 75 \\ 75 & 0 & 75 & 0 \\ 75 & 75 & 0 & 75 \\ 75 & 0 & 75 & 0 \end{bmatrix}. \tag{7.18}$$

1. Solve an optimal routing problem for the pipe model presented in Eqs. (7.2a)–(7.2f) and obtain $r_P$. The traffic matrix, $\mathbf{T} = \{t_{pq}\}$, where $t_{pq}$ is the traffic demand from node $p$ to node $q$, is given by

$$T = \begin{bmatrix} 0 & 35 & 35 & 35 \\ 35 & 0 & 35 & 35 \\ 35 & 35 & 0 & 35 \\ 35 & 35 & 35 & 0 \end{bmatrix}. \tag{7.19}$$

2. Solve an optimal routing problem for the hose model presented in Eqs. (7.9a)–(7.9h), and obtain $r_H$. Set $\alpha_p = \sum_q t_{pq}$ and $\beta_q = \sum_p t_{pq}$ using $\mathbf{T}$ defined in Eq. (7.19).

3. Solve an optimal routing problem for the HSDT model presented in Eqs. (7.13a)–(7.13h), and obtain $r_{HSDT1}$. Set $\alpha_p = \sum_q t_{pq}$ and $\beta_q = \sum_p t_{pq}$ using $\mathbf{T}$ in Eq. (7.19). Set $\delta_{pq}$ and $\gamma_{pq}$ so that the 25% upper-bound margin and 20% lower-bound margin of $t_{pq}$ in Eq. (7.19) can be kept, respectively.

4. Solve the same optimal routing problem for the HSDT model as the above one by changing only the margin condition to 50% upper-bound margin and 50% lower-bound margin of $t_{pq}$, and obtain $r_{HSDT2}$.

5. Solve an optimal routing problem for the HLT model presented in Eqs. (7.17a)-(7.17h), and obtain $r_{HLT}$. Set $\alpha_p = \sum_q t_{pq}$, $\beta_q = \sum_p t_{pq}$, and $y_{ij} = \sum_{p,q \in Q} a_{ij}^{pq} t_{pq}$ using $\mathbf{T}$ defined in Eq. (7.19) and $a^{pq}$, which gives the routing information in the first phase of HLT. In the first phase, the shortest path routing is used, where the cost of link $(i, j)$, $d_{ij}$, is set to 1.

6. Compare $r_P$, $r_H$, $r_{HSDT1}$, $r_{HSDT2}$, and $r_{HLT}$ obtained above.

## Exercise 7.2

Consider the network in Figure 7.7. The capacity matrix, $\mathbf{C} = \{c_{ij}\}$, where $c_{ij}$ is the capacity of link $(i, j)$, is given by

$$C = \begin{bmatrix} 0 & 100 & 100 & 100 & 0 & 100 \\ 100 & 0 & 100 & 100 & 100 & 100 \\ 100 & 100 & 0 & 0 & 100 & 0 \\ 100 & 100 & 0 & 0 & 100 & 100 \\ 0 & 100 & 100 & 100 & 0 & 100 \\ 0 & 100 & 0 & 100 & 100 & 0 \end{bmatrix}. \tag{7.20}$$

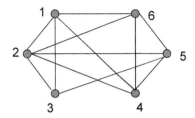

Figure 7.7: Network model.

1. Solve an optimal routing problem for the pipe model presented in Eqs. (7.2a)–(7.2f) and obtain $r_P$. The traffic matrix, $\mathbf{T} = \{t_{pq}\}$, where $t_{pq}$ is the traffic demand from node $p$ to node $q$, is given by

$$
\mathbf{T} =
\begin{bmatrix}
0 & 35 & 35 & 35 & 35 & 35 \\
35 & 0 & 35 & 35 & 35 & 35 \\
35 & 35 & 0 & 35 & 35 & 35 \\
35 & 35 & 35 & 0 & 35 & 35 \\
35 & 35 & 35 & 35 & 0 & 35 \\
35 & 35 & 35 & 35 & 35 & 0
\end{bmatrix}.
\tag{7.21}
$$

2. Solve an optimal routing problem for the hose model presented in Eqs. (7.9a)–(7.9h), and obtain $r_H$. Set $\alpha_p = \sum_q t_{pq}$ and $\beta_q = \sum_p t_{pq}$ using $\mathbf{T}$ defined in Eq. (7.21).

3. Solve an optimal routing problem for the HSDT model presented in Eqs. (7.13a)–(7.13h), and obtain $r_{HSDT1}$. Set $\alpha_p = \sum_q t_{pq}$ and $\beta_q = \sum_p t_{pq}$ using $\mathbf{T}$ in Eq. (7.21). Set $\delta_{pq}$ and $\gamma_{pq}$ so that a 25% upper-bound margin and 20% lower-bound margin of $t_{pq}$ in Eq. (7.21) can be kept, respectively.

4. Solve the same optimal routing problem for the HSDT model as the above one by changing only the margin condition to a 50% upper-bound margin and 50% lower-bound margin of $t_{pq}$, and obtain $r_{HSDT2}$.

5. Solve an optimal routing problem for the HLT model presented in Eqs. (7.17a)–(7.17h), and obtain $r_{HLT}$. Set $\alpha_p = \sum_q t_{pq}$, $\beta_q = \sum_p t_{pq}$, and $y_{ij} = \sum_{p,q \in Q} a_{ij}^{pq} t_{pq}$ using $\mathbf{T}$ defined in Eq. (7.21) and $a^{pq}$, which gives the routing information in the first phase of HLT. In the first phase,

the shortest path routing is used, where the cost of link $(i, j)$, $d_{ij}$, is set to 1.

6. Compare $r_P$, $r_H$, $r_{HSDT1}$, $r_{HSDT2}$, and $r_{HLT}$ obtained above.

## Exercise 7.3

Prove Property 2 presented in Section 7.4.

## Exercise 7.4

Prove Property 3 presented in Section 7.5.

# Bibliography

[1] R. Zhang-Shen and N. McKeown, "Designing a fault-tolerant network using valiant load-balancing," *IEEE Infocom 2008*, April 2008.

[2] A. Juttner, I. Szabo, A. Szentesi, "On bandwidth efficiency of the hose resource management model in virtual private networks," *IEEE Infocom 2003*, pp. 386–395, Mar./Apr. 2003.

[3] N. G. Duffield, P. Goyal, A. Greenberg, P. Mishra, K. K. Ramakrishnan, and J. E. van der Merwe, "Resource management with hoses: Point-to-cloud services for virtual private networks," *IEEE/ACM Trans. on Networking*, vol. 10, no. 5, pp. 679–692, Oct. 2002.

[4] A. Kumar, R. Rastogi, A. Silberschatz, and B. Yener, "Algorithms for provisioning virtual private networks in the hose model," *The 2001 Conference on Applications, Technologies, Architectures, and Protocols for Computer Communications*, pp. 135–146, 2001.

[5] Y. Wang and Z. Wang, "Explicit routing algorithms for Internet traffic engineering," *IEEE International Conference on Computer Communications and Networks (ICCCN)*, 1999.

[6] J. Chu and C. Lea, "Optimal link weights for maximizing QoS traffic," *IEEE ICC 2007*, pp. 610–615, 2007.

[7] J. Chu and C. Lea, "Optimal link weights for IP-based networks supporting hose-model VPNs," *IEEE/ACM Trans. Networking*, vol. 17, no. 3, pp. 778–788, June 2009.

[8] M. Kodialam, T. V. Lakshman, J. B. Orlin, and S. Sengupta, "Pre-configuring IP-over-optical networks to handle router failures and unpredictable traffic," *IEEE Infocom 2006*, Apr. 2006.

[9] M. Kodialam, T. V. Lakshman, J. B. Orlin, and S. Sengupta, "Oblivious Routing of Highly Variable Traffic in Service Overlays and IP Backbones," *IEEE/ACM Trans. Networking*, vol. 17, no. 2, pp. 459–472, Apr. 2009.

[10] D. Applegate and E. Cohen, "Making intra-domain routing robust to changing and uncertain traffic demands: Understanding fundamental tradeoffs," *Proc. of SIGCOMM'03*, 2003.

[11] D. Applegate and E. Cohen, "Making routing robust to changing traffic demands: Algorithms and evaluation," *IEE/ACM Trans. Networking*, vol. 14, no. 6, pp. 1193–1206, 2006.

[12] B. Towles and W. J. Dally, "Worst-case traffic for oblivous routing functions," *IEEE Computer Architecture Letters*, vol. 1, no.1, 2002.

[13] M. Bienkowski, M. Korzeniowski, and H. Räcke, "A practical algorithm for constructing oblivious routing schemes," *Proc. of SPAA'03*, 2003.

[14] T. Mikoshi, T. Takenaka, T. Fujiwara, E. Oki, and K. Shiomoto, "Optimization of input admissible traffic flow guaranteeing QoS," *IEEE Commun. Let.*, vol. 13, no. 1, pp. 49–51, Jan. 2009.

[15] A. Medina, N. Taft, K. Salamatian, S. Bhattacharyya, and C. Diot, "Traffic matrix estimation: Existing techniques and new directions," *ACM SIGCOMM 2002*, pp. 161–174, Aug. 2002.

[16] Y. Zhang, M. Roughan, N. Duffield, and A. Greenberg, "Fast accurate computation of large-scale IP traffic matrices from link loads," *ACM SIGMETRICS 2003*, pp. 206–217, June 2003.

[17] A. Nucci, R. Cruz, N. Taft, and C. Diot, "Design of IGP link weight changes for estimation of traffic matrices," *INFOCOM 2004*, vol. 4, pp. 2341–2351, Mar. 2004.

[18] Y. Ohsita, T. Miyamura, S. Arakawa, S. Ata, E. Oki, K. Shiomoto, and M. Murata, "Gradually reconfiguring virtual network topologies based on estimated traffic matrices," *INFOCOM 2007*, pp. 2511–2515, May 2007.

[19] E. Oki and A. Iwaki, "Performance Comparisons of Optimal Routing by Pipe, Hose, and Intermediate Models," *IEEE Sarnoff 2009*, Mar.–Apr. 2009.

[20] E. Oki and A. Iwaki, "Performance of optimal routing by pipe, hose, and intermediate models," *IEICE Trans. Commun.*, vol. E93-B, no. 5, pp. 1180–1189, May 2010.

[21] Y. Kitahara and E. Oki, "Optimal routing strategy by hose model with link-traffic bounds," *IEEE Globecom 2011*, Dec. 2011.

# Chapter 8

# IP routing

This chapter describes several problems raised by route selection in an Internet Protocol (IP) network. In an IP network, an IP packet with the destination address is transmitted by way of transit node(s) to the destination node. When a node receives an IP packet, it determines the next hop node to which the packet should be transmitted by referring to its routing table made by IP routing protocols. By repeating this process, a packet reaches its destination.

## 8.1 Routing protocol

IP routing protocols are used to dynamically update routing tables, which indicate the next hop node according to the destinations of incoming packets. A routing protocol requires nodes exchange network information such as links and nodes in the network.

The concept of the routing table is similar to that of road signs seen when driving your car to a destination, as shown in Figure 8.1. In Figure 8.1(a), the driver wants to go to the harbor. When the car enters the junction, it turns to the right by referring to the road sign. The driver does not have to know the entire route to the destination in advance. Each time the car enters a junction, the road sign tells the driver the suitable choice. As a result, the driver is able to reach the harbor. In Figure 8.1(b), the destination of the incoming packet is 123.10.21.50. The node forwards the packet to output port 3, which is associated with the next hop, by referring to the routing table. In the same way as the car, each packet reaches its destination.

IP routing protocols are categorized into several types, including a link-state protocol [1,2], a distance-vector protocol [3,4], and a path-vector protocol [5]. This chapter focuses on the link-state protocol, where an optimization problem is considered for routing. Some popular link-state protocols are the Open Shortest Path First (OSPF) [1] and Intermediate System to Intermediate System (IS-IS) [2]. In a link-state protocol, each node advertises link

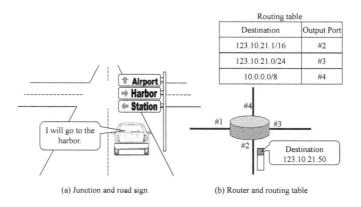

Figure 8.1: Road map and IP routing table.

states, including topology information about nodes, links, and their costs to their neighbor nodes. The neighbor nodes that receive the advertised link states advertise them to their neighbor nodes. This iterative advertisement process, called *flooding*, makes all the nodes share the same topology information, as shown in Figure 8.2. Each node is able to compute the shortest path to each destination using costs, or *weights*, each of which is associated with a different link.

## 8.2   Link weights and routing

Determining the link weights in the network means determining the routing based on shortest path routing. A simple weight setting policy is to make the link weight inversely proportional to its capacity. This policy, which is implemented in commercial routers, makes it easy for network providers to configure routers to avoid network congestion. Traffic tends to avoid links with small capacity and instead use those with large capacity. However, as this policy does not take traffic demand or network topology into consideration, it may degrade the routing performance. Therefore, to achieve good routing performance, the link weights should be determined by considering the traffic demand and topology.

Let us consider the network in Figure 8.3. The weight associated with each link is set to 1. Assume that traffic demands are sent from nodes 1, 2,

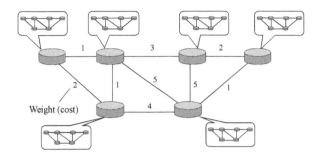

Figure 8.2: Link-state routing protocol.

3, and 5 to node 7 in the network. Each has traffic volume of 1. Each node has the network topology with all weights, computes the shortest paths, and configures the routing table. As shown in Figure 8.4, the shortest-path routes are determined as $1 \to 4 \to 7$, $2 \to 4 \to 7$, $3 \to 4 \to 7$, and $5 \to 6 \to 7$. As a result, the traffic volume passing through link $(4, 7)$ is 3, and the link is congested.

To avoid the congestion, the weight of link $(1, 4)$ is changed from 1 to 3, as shown in Figure 8.5. As shown in Figure 8.6, the updated shortest-path routes are determined as $1 \to 5 \to 6 \to 7$, $2 \to 4 \to 7$, $3 \to 4 \to 7$, and $5 \to 6 \to 7$. The congestion at link $(4, 7)$ is relaxed. The traffic volume passing through the link is changed from 3 to 2 because $1 \to 4 \to 7$, the previous route, is changed to $1 \to 5 \to 6 \to 7$ to avoid the link.

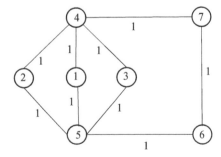

Figure 8.3: Network model and link weights (example 1).

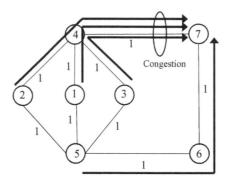

Figure 8.4: Link weights and route selection (example 1).

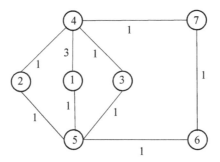

Figure 8.5: Network model and link weights (example 2).

### 8.2.1  Tabu search

Let us consider finding an optimal set of link weights to minimize the network congestion ratio, which is the maximum link utilization rate of all links in the network. A straightforward approach is to compute the network congestion ratios for every possible set of link weights to find the optimal set that minimizes the network congestion ratio. However, the computation time complexity of this approach is $O(x^L)$, where $x$ is the higher limit of weights and $L$ is the number of links in the network. To reduce the computation-time complexity,

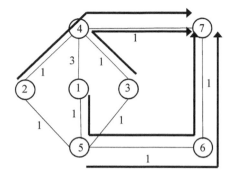

Figure 8.6: Link weights and route selection (example 2).

Fortz et al. presented a heuristic algorithm based on tabu search [6,7]. Buriol et al. presented a genetic algorithm with a local improvement procedure [8]. A fast heuristic algorithm was also developed by Reichert and Magedanz [9]. These optimization algorithms yield nearly optimal sets of link weights in a practical manner.

The tabu search (TS) methodology [10] is explained here. TS is an iterative procedure designed to solve optimization problems. It is a based on selected concepts that unite the fields of artificial intelligence and optimization. It has been applied to a wide range of problems such as job scheduling, graph coloring, and network planning. It is considered an alternative to techniques such as simulated annealing and genetic algorithms. There are several studies that use TS to search for an optimal OSPF link-weight set [6,7,11]. All these studies confirmed that TS was useful.

TS conducts a guided exploration of the space of admissible solutions and keeps a record of all solutions evaluated along the way, which is called a tabu list. The exploration starts from an initial solution. In the evaluation of admissible solutions, any solution existing in the tabu list is not selected even if it has lower cost than current one. If no admissible solution has lower cost than the current solution, TS permits selection of the admissible solution with high cost. When a stop criterion is satisfied, the algorithm returns the best visited solution. To move from one solution to the next, TS explores the neighborhood of the last solution visited (referred to as the current solution). It generates a neighbor solution by applying a transformation, called a move, on the current solution. The set of all admissible moves uniquely defines the neighborhood of the current solution. At each iteration of the TS algorithm, all solutions in the neighborhood are evaluated and the best is selected as the new current solution.

The TS algorithm to find the optimal link-weight set is described below. At the beginning, an empty tabu list, that is, no link-weight set, is created.

Step 1: *Set initial candidate*: As an initial candidate, a set of link weights randomly generated that is not in the tabu list is considered the initial solution at this iteration. The solution for each iteration is called $W_{itr}$, and the final result is called the solution, $W_{opt}$. The initial candidate is set in $W_{itr}$, and $W_{opt} = W_{itr}$ is set. The network congestion ratio is obtained based on shortest-path routing with the initial weight set.

Step 2: *Find highest congested link*: Find the highest congested link and the topology for which the congestion occurs.

Step 3: *Move to next candidate*: At each time, the previous data of the highest congested link is used. The next candidate is created by increasing the link weight of the marked link. The link weight of the marked link is increased by the minimum value that changes at least one route passing through the link. The minimum value is identified by incrementing the weight of the marked link, until at least one route passing through that link is changed. This, therefore, decreases the congestion of the marked link. The updated candidate is inserted in the tabu list. If the value exceeds the upper limit of feasible link weight, go to Step 6.

Step 4: *Evaluation of candidate*: The updated link-weight set obtained at Step 3 is evaluated by computing the network congestion. If the network congestion ratio with the updated set is lower than that of $W_{itr}$ set, the weight set is set as $W_{itr}$ and the next candidate, and Step 2 is reentered. Otherwise, return to Step 2 to find out the highest congested link again for the link-weight set obtained at Step 3 and iterate steps 3 and 4. If the number of loops from Step 2 to Step 4 exceeds a predetermined number, $C_{max}$, go to Step 5.

Step 5: *Keep optimum solution*: If the congestion ratio of $W_{itr}$ is lower than that of $W_{opt}$, set $W_{opt} = W_{itr}$.

Step 6: *Continue iteration with stop criteria*: Go to Step 1 for the next iteration, unless the number of iterations does not exceed a fixed predetermined value, $I_{max}$. $I_{max}$ depends on network size, the allowable computational time, and the quality of solution desired. Otherwise, the search procedure stops, and $W_{opt}$ is the solution.

Figures 8.7–8.11 show an example of the TS algorithm that determines a suitable set of link weights. In Figure 8.7, a set of link weights is given as the initial condition. Consider that traffic demands from nodes 1, 2, 3, and 7 to node 6 exist in the network. Each has a traffic volume of 1. Assume that all link capacities are the same. Minimizing the link utilization is equivalent to minimizing the traffic volume passing through the link in this example.

Therefore, we consider minimizing the maximum traffic volume on the link in the network by the TS search. The shortest-path routes are determined as $1 \rightarrow 4 \rightarrow 6$, $2 \rightarrow 4 \rightarrow 6$, $3 \rightarrow 4 \rightarrow 6$, and $7 \rightarrow 8 \rightarrow 6$. As a result, the traffic volume passing through link $(4, 6)$, which is the highest congested link, is 3.

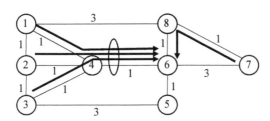

Figure 8.7: Example of tabu search (initial state).

In Figure 8.8, to decrease the traffic volume passing through link $(4, 6)$, the link weight is increased by the minimum value that changes at least one route passing through the link to decrease the congestion. The link weight is changed from 1 to 4. The highest congested link switches from $(4, 6)$ to $(8, 6)$. The traffic volume on link $(8, 6)$, which is now the highest one in the network, becomes 2.

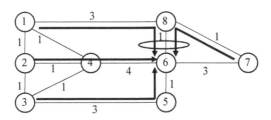

Figure 8.8: Example of tabu search (after first change).

In Figure 8.9, to decrease the traffic volume passing through link $(8, 6)$, the link weight is changed from 1 to 3. The highest congested link is switched from $(8, 6)$ to $(4, 6)$. The traffic volume on link $(4, 6)$, which is now the highest one in the network, becomes 2.

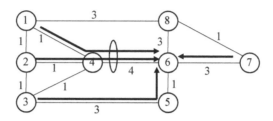

Figure 8.9: Example of tabu search (after second change).

In Figure 8.10, to decrease the traffic volume passing through link $(4, 6)$, the link weight is changed from 4 to 5. The highest congested link is switched from $(4, 6)$ to $(3, 5)$ and $(5, 6)$. The traffic volume on links $(3, 5)$ and $(5, 6)$, which is now the highest one in the network, becomes 2.

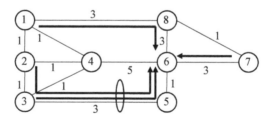

Figure 8.10: Example of tabu search (after third change).

In Figure 8.11, as links $(3, 5)$ and $(5, 6)$ are the highest congested link, $(5, 6)$ is randomly selected. To decrease the traffic volume passing through link $(5, 6)$, the link weight is changed from 1 to 3. As a result, all the traffic flows are evenly distributed and the highest traffic volume on each link becomes 1.

This example shows a case with an initial condition for a set of link weights. In a practical case, multiple initial conditions are used, and the most suitable set in terms of lowering the network congestion is selected.

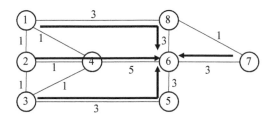

Figure 8.11: Example of tabu search (after fourth change).

# 8.3 Preventive start-time optimization (PSO)

## 8.3.1 Three policies to determine link weights

To determine a set of link weights, there are three optimization policies: Start-time Optimization (SO), Runtime Optimization (RO), and Preventive Start-time Optimization (PSO).

When the network topology and traffic matrix are given, SO can determine the optimal set of link weights once at the beginning of network operation. Unfortunately, SO is weak against network failure, for example, a link failure. The most crucial point of SO is that it considers the link weight assignment problem as a static problem and ignores network dynamic changes at run-time. In practice, one of the main challenges for network operators is to deal with link failures, which occur on a daily basis in large IP backbones [12]. Because a link failure will trigger the rerouting of some active paths, the SO-generated weight set is no longer optimal. This can cause unexpected network congestion. Nucci at el. also referred to this problem in [11] and introduced the concept of providing robustness without changing weights during short-lived link failure events. However, they did not present how to determine the set of link weights at the start time so as to secure this robustness.

The weakness of SO can be overcome by computing a new optimal set of link weights whenever the topology is changed. This approach, called RO, obviously provides the best routing performance after each link failure. However, changing link weights during a failure may not be practical for two reasons. First, the updated weights must be flooded to every router in the network, and every router must then re-compute its minimum cost path to every other router. This can lead to considerable instability in the network. Meanwhile, IP packets may arrive in disorder, and the performance of Transport Control Protocol (TCP) connections may be degraded [6, 7]. The more often link weights are changed, the longer the network takes to achieve stability as packets are sent back and forth between routers to achieve the very divergent processes of updating the routing table and calculating the shortest paths based on the new link weights. Large networks exhibit more pronounced network instability when link failure occurs. The second reason is related to the short-lived nature of most link failures. In [12], inter-PoP link failures over a 4-month period were examined, and it was found that 80% of the failures lasted less than 10 minutes, and 50% of the failures lasted less than 1 minute. They thus define transient failures as those failures that last for not more than 10 minutes. Transient failures can create rapid congestion that is harmful to the network [13]. However, they leave the human operator with insufficient time to reassign link weights before the failed link is restored. The study in [12] also examined the frequency of single-link and multiple-link failures. They observed that more than 70% of transient failures are single-link failures. Therefore, it seems reasonable to target the one-time configuration of link weights that can handle any link failure.

PSO determines, at the start time, a suitable set of link weights that can handle any possible link failure scenario preventively [14]. Only a single link failure is considered in [14]. However, the concept of PSO is easily extended to multiple link failures. The objective is to determine, at the start time, the most appropriate set of link weights that can avoid both unexpected network congestion and network instability, the drawbacks of SO and RO, respectively, regardless of which link fails. PSO considers all possible link failure scenarios at start time in order to determine a suitable set of link weights. Numerical results showed that PSO is able to reduce the worst-case network congestion ratio, compared to that of SO, while it also avoids the runtime changes of link weights, which would otherwise trigger network instability.

To determine a suitable set of link weights based on the PSO policy, there are two PSO-based algorithms: PSO-L (limited rage of candidates) [14] and PSO-W (wide range of candidates) [15]. In the following sections, the PSO model is defined and PSO-L and PSO-W are explained.

### 8.3.2   PSO model

The network is represented as directed graph $G(V, E)$, where $V$ is the set of nodes and $E$ is the set of links. $v \in V$, where $v = 1, 2, \cdots, N$, indicates an individual node and $e$, where $e = 1, 2, \cdots, L$, indicates a bidirectional individual link. $N$ is the number of nodes and $L$ is the number of links in the network. We consider only single link failure in this work, as the probability of concurrent multiple link failure is much less than that of single link failure. $F$ is the set of link failure indices $l$, where $l = 0, 1, 2, \cdots, L$ and $F = E \cup \{0\}$. The number of elements in $F$ is $|F| = L + 1$. $l = 0$ indicates no link failure and $l \ (\neq 0)$ indicates the failure of link $e = l \in E$. $G_l$ denotes $G$ that has no link $e = l \ (\neq 0)$ because of link failure[1]. $c_e$ is the capacity of $e \in E$. The traffic

---

[1] $G_0 = G$ as $l = 0$ indicates no failure.

volume of passing through $e$ is denoted as $u_e$. $T = \{t_{pq}\}$ is the traffic matrix, where $t_{pq}$ is the traffic demand from source node $p$ to destination node $q$.

The network congestion ratio $r$ refers to the maximum value of all link utilization ratios in the network. $r$ is defined by

$$r = \max_{e \in E} \frac{u_e}{c_e}, \tag{8.1}$$

where $0 \le r \le 1$. Traffic volume $\frac{1-r}{r} t_{pq}$ is the greatest volume that can be added to the existing traffic volume of $d_{pq}$ for any pair of source node $p$ and destination node $q$ such that the traffic volume passing though any $e$ does not exceed $c_e$ under the condition that routing is not changed. After $\frac{1-r}{r} t_{pq}$ is added to $d_{pq}$, the total traffic volume becomes $\frac{1}{r} t_{pq}$ and the updated network congestion becomes 1, which is the upper limit. Maximizing the additional traffic volume of $\frac{1-r}{r} t_{pq}$ is equivalent to minimizing $r$ [16].

$W = \{w_e\}$ is the $L \times 1$ link weight matrix of network $G$, where $w_e$ is the weight of link $e$. $W_{cand}$ is the set of candidate $W$ for which we are calculating the worst-case congestion. $r(W, l)$ is a function that returns the congestion ratio defined in Eq. (8.1) for $G_l$ according to OSPF-based shortest path routing using the link weights in $W$. $R(W)$ refers to the worst-case congestion ratio in $W$ among all link failure scenarios $l \in F$. $R(W)$ is defined by

$$R(W) = \max_{l \in F} r(W, l). \tag{8.2}$$

Our target is to find the most appropriate set of link weights, $W_{PSO}$, for network $G$ that minimizes $R(W)$ defined in Eq. (8.2) over link failure index $l \in F$. $W_{PSO}$ is defined by

$$W_{PSO} = \arg \min_{W \in W_{cand}} R(W), \tag{8.3}$$

where $W_{cand}$ is all possible link-weight set candidates. The network congestion ratio achieved using $W_{PSO}$ is $R(W_{PSO})$; it represents the upper bound of congestion for any single link failure scenario in the network.

## 8.3.3   PSO-L

### 8.3.3.1   Overview of PSO-L

PSO-L determines $W_{PSO}$ out of link-weight set candidates in $W_{cand}$, which is limited by PSO-L, in Eq. 8.3. For PSO-L, we use the definitions stated below: $W_l^* = \{w_1, w_2, \cdots, w_{l-1}, w_{l+1}, \cdots, w_L\}$ is an $(L-1) \times 1$ link-weight matrix that is optimized for network $G_l$. $W_l^k = \{w_1, w_2, \cdots, w_{l-1}, w_l^k, w_{l+1}, \cdots, w_L\}$ is an $L \times 1$ link-weight matrix, where $w_l^k$ is obtained by the optimization of $G_k$ ($k \ne l \in F$) and the other link weights are set using $W_l^*$. $W_l^k$ is represented by

$$W_l^k = W_l^* \cup \{w_l^k\}. \tag{8.4}$$

$W_l$, which is an $L \times 1$ link-weight matrix, refers to $W_l^k$ defined in Eq. (8.4) that minimizes $R(W_l^k)$ defined in Eq. (8.2) over $k \in F$. $W_l$ is defined by

$$W_l = \arg \min_{k(\neq l) \in F} R(W_l^k). \tag{8.5}$$

The target of PSO is to find the most appropriate set of link weights, $W_{PSO}$, for network $G$ that minimizes $R(W_l)$ defined in Eq. (8.2) over link failure index $l \in F$. In PSO-L, $W_{PSO}$ is defined by

$$W_{PSO} = \arg \min_{l \in F} R(W_l). \tag{8.6}$$

The procedure used by PSO-L to obtain $W_{PSO}$, defined by Eq. (8.6), is divided into three steps as follows:

Step 1: Get $W_l$ for all $l \in F$.

Step 2: Compute $R(W_l)$ for all $W_l$, where $l \in F$, using Eq. (8.2).

Step 3: Get $W_{PSO}$ using Eq. (8.6).

Step 1 is also divided into three sub-steps to obtain $W_l$, as follows:

Step 1a: Compute $W_l^k$ for all $k \in F$, using Eq. (8.4).

Step 1b: Compute $R(W_l^k)$ for all $k \in F$, using Eq. (8.2).

Step 1c: Get $W_l$ that minimizes $R(W_l^k)$ for all $k \in F$, using Eq. (8.4).

If there is more than one equivalent value of $W_l^k$ that minimizes $R(W_l^k)$, one $W_l^k$ is determined based on the following policy. If $W_l^k$ with $k = 0$ is included, it is selected. Otherwise, one $W_l^k$ is randomly selected out of multiple candidates.

### 8.3.3.2   Examples of PSO-L

Figure 8.12 is an example network with four nodes and five links, which are denoted $e = 1$, 2, 3, 4, and 5. We consider the no link failure case and all cases of possible link failure, from $l = 0$ to $l = 5$. Consider the case of no link failure, $l = 0$, as shown in Figure 8.12(a). The optimal set of link weights with $l = 0$ is denoted as $W_0$.

Next, consider $l = 1$, where link $e = 1$ has failed, as shown in Figure 8.12(b). $W_1^* = \{w_2, w_3, w_4, w_5\}$, where $w_1$ is not included, is obtained by considering the network topology with $l = 1$. $w_1$ must be determined to get $W_1$. To get $w_1$, the topology with link failure $k \in F(\neq 1)$ is considered, where $w_1^k$ is obtained. Then, five sets of link weights denoted as $W_1^k = \{w_1^k, w_2, w_3, w_4, w_5\}$ are obtained. Using $W_1^k$, we obtain the network congestion ratios with link failure $l$, denoted as $r(W_1^k, l)$, as shown in Table 8.1. Next, we obtain the worst-case congestion ratio over $l \in F$,

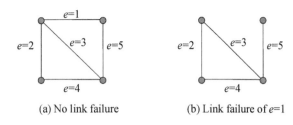

(a) No link failure          (b) Link failure of $e=1$

Figure 8.12: Network model.

Table 8.1: $r(W_1^k, l)$ for each link failure scenario.

| Link failure index $l$ | $W_1^0$ | $W_1^2$ | $W_1^3$ | $W_1^4$ | $W_1^5$ |
|---|---|---|---|---|---|
| 0 | 0.05 | 0.05 | 0.05 | 0.05 | 0.06 |
| 1 | 0.04 | 0.04 | 0.04 | 0.04 | 0.04 |
| 2 | 0.11 | 0.10 | 0.11 | 0.13 | 0.13 |
| 3 | 0.08 | 0.04 | 0.04 | 0.07 | 0.06 |
| 4 | 0.07 | 0.05 | 0.06 | 0.05 | 0.06 |
| 5 | 0.06 | 0.08 | 0.08 | 0.06 | 0.06 |
| $R(W_1^k)$ | 0.11 | 0.10 | 0.11 | 0.13 | 0.13 |

(©2011 IEICE, Ref. [15], P. 1967.)

or $R(W_1^k) = \max_{l \in F} r(W_1^k, l)$. In the last row of Table 8.1, we find that the worst-case congestion ratio over $l$ is minimum for $W_1^2$. As a result, $W_1 = W_1^2 = \{w_1^2, w_2, w_3, w_4, w_5\}$ is determined.

In the same way as $W_1$, we obtain $W_2$, $W_3$, $W_4$, and $W_5$ for each link failure scenario. Using $W_l$, where $l \in F$, we obtain the network congestion ratios with link failure $l' \in F$ denoted as $r(W_l, l')$, as shown in Table 8.2. Next, we obtain the maximum congestion ratio over $l' \in F$, or $R(W_l) = \max_{l' \in F} r(W_l, l')$. In the last row of Table 8.2, we find that the maximum congestion ratio over $l'$ is minimum for $W_2$.

### 8.3.3.3   Problem of PSO-L

Our desired goal is to determine a set to *minimize* the worst case congestion. However, PSO-L is not guaranteed to obtain an optimal set that minimizes the worst-case congestion. Numerical results presented in [14] indicate that PSO-L is able to reduce the worst-case network congestion ratio, but it is not verified that the obtained set minimizes the worst-case network congestion ratio.

In PSO-L, only $W_l$ for $l = 0, \ldots, L$, each of which is optimized set for $G_l$, is

Table 8.2: $r(W_l, l')$ for each link failure scenario and $R(W_l)$

| Link failure index $l'$ | $W_0$ | $W_1$ | $W_2$ | $W_3$ | $W_4$ | $W_5$ |
|---|---|---|---|---|---|---|
| 0 | 0.04 | 0.05 | 0.05 | 0.13 | 0.04 | 0.09 |
| 1 | 0.04 | 0.04 | 0.06 | 0.11 | 0.04 | 0.11 |
| 2 | 0.11 | 0.10 | 0.07 | 0.13 | 0.11 | 0.09 |
| 3 | 0.04 | 0.04 | 0.05 | 0.04 | 0.04 | 0.06 |
| 4 | 0.05 | 0.05 | 0.09 | 0.13 | 0.05 | 0.13 |
| 5 | 0.06 | 0.08 | 0.06 | 0.06 | 0.06 | 0.06 |
| $R(W_l)$ | 0.11 | 0.10 | 0.09 | 0.13 | 0.11 | 0.13 |

(©2011 IEICE, Ref. [15], P. 1967.)

considered a candidate weight set. However, a link weight set, $\Omega$, which is not optimized set for any $G_l$, may give a lower worst-case congestion ratio than those of any $W_l$. In other words, $\Omega$ that satisfies $R(\Omega) \leq R(W_l)$ may exist. Table 8.3 shows a possible numerical example of congestion ratios for $W_l$ for $l = 0, \ldots, L$ and $\Omega$, assuming that $\Omega$ that satisfies $R(\Omega) \leq R(W_l)$ exists, while $\Omega$ is not considered as a candidate in PSO-L. Note that the same network that is employed in Section 8.3.3.2, as shown in Figure 8.12(a), is used.

Table 8.3: $r(W_l, l')$ and $r(\Omega, l')$ for each link failure scenario and their worst congestion ratios, where $R(\Omega) \leq R(W_l)$.

| Link failure index $l'$ | $W_0$ | $W_1$ | $W_2$ | $W_3$ | $W_4$ | $W_5$ | $\Omega$ |
|---|---|---|---|---|---|---|---|
| 0 | 0.04 | 0.05 | 0.05 | 0.13 | 0.04 | 0.09 | 0.05 |
| 1 | 0.04 | 0.04 | 0.06 | 0.11 | 0.04 | 0.11 | 0.05 |
| 2 | 0.11 | 0.10 | 0.07 | 0.13 | 0.11 | 0.09 | 0.08 |
| 3 | 0.04 | 0.04 | 0.05 | 0.04 | 0.04 | 0.06 | 0.06 |
| 4 | 0.05 | 0.05 | 0.09 | 0.13 | 0.05 | 0.13 | 0.07 |
| 5 | 0.06 | 0.08 | 0.06 | 0.06 | 0.06 | 0.06 | 0.06 |
| $R(W_l)$ | 0.11 | 0.10 | 0.09 | 0.13 | 0.11 | 0.13 | 0.08 |

(©2011 IEICE, Ref. [15], P. 1968.)

In Table 8.3, $W_2$ is the solution of PSO-L, which satisfies Eq. (8.5). Note that at least one of $r(W_l, l')$ is always equal to or lower than $r(\Omega, l')$ for any link failure scenario $l'$. However, as $R(\Omega)$ is lower than $R(W_2)$, $W_2$ does not give the minimum worst-case congestion ratio. It is not guaranteed that PSO-L always gives the *optimal* solution to minimize the worst-case congestion ratio.

## 8.3.4 PSO-W

### 8.3.4.1 Overview of PSO-W

PSO-W uses all $W \in W_{cand}$ to calculate $R(W)$, using Eq. (8.2) and find the optimal $W$ for which $R(W)$ is minimum [15]. That ensures minimization of the worst-case congestion ratio. This procedure defined by Eq. (8.3) is divided into two steps as follows:

Step 1: Compute $R(W)$ for all $W \in W_{cand}$, using Eq. (8.2).

Step 2: Get $W_{PSO}$ using Eq. (8.3).

The issue of immediate concern is the high number of weight matrix candidates in $W_{cand}$. It is obvious that the more the candidates are, the more likely it is of getting the appropriate matrix. If the allowable upper limit of a link weight is $x$, the number of possible candidates is $x^L$. To reduce the time complexity, PSO-L considers only the optimized link set for possible topologies brought from single link failures, that is, $L + 1$. This smaller number of candidates decreases the possibility of getting the optimal worst-case performance. However, the PSO-L candidates may not include link-weight sets that are not optimized for any topology due to network failure, but that may give lower worst-case congestion and thus the best performance.

## 8.3.5 PSO-W algorithm based on tabu search

PSO-W finds a more suitable link-weight set in a wide range of candidates by adopting the TS-based algorithm in a hermitic manner than PSO-L. The PSO-W algorithm is described below. At the beginning, a tabu list, where no link-weight set exits, is set.

### 8.3.5.1 PSO-W algorithm

*Objective function*: Find weight set $W_{PSO}$ that meets Eq. (8.3), the set that gives the minimum worst-case congestion.

Step 1: *Set initial candidate*: As an initial candidate, a set of link weights randomly generated that is not in the tabu list is considered the initial solution at this iteration. The solution for each iteration is called $W_{itr}$, and the final result is called the solution, $W_{opt}$. The initial candidate is set in $W_{itr}$, and $W_{opt} = W_{itr}$ is set. The network congestion ratio is obtained based on the shortest-path routing with the initial weight set.

Step 2: *Find worst-case highest congested link*: The worst-case congestion is computed using Eq. (8.2) against all topologies created by single link failure. Find the worst-case highest congested link and the topology for which the congestion occurs.

Step 3: *Move to next candidate*: At each time, the previous data of the worst-case highest congested link is used. The next candidate is created by increasing the link weight of the marked link. The link weight of the marked link is increased by the minimum value that changes at least one route passing through the link. The minimum value is searched by incrementing the weight of the marked link, until at least one route passing through that link is changed. This, therefore, decreases the congestion of the marked link. The updated candidate is inserted in the tabu list. If the value exceeds the upper limit of feasible link weight, go to Step 6.

Step 4: *Evaluation of candidate*: The updated link-weight set obtained at Step 3 is evaluated by computing the network congestion. It considers all single link failure topologies to compute the worst-case congestion. If the network congestion ratio with the updated set is lower than that of the $W_{itr}$ set, the weight set is set as $W_{itr}$ and the next candidate, and Step 2 is reentered. Otherwise return to Step 2 to find out the worst-case highest congested link again for the link-weight set obtained at Step 3 and iterate Steps 3 and 4. If the number of loops from Step 2 to Step 4 exceeds the predetermined number, $C_{max}$, go to Step 5.

Step 5: *Keep optimum solution*: If the congestion ratio of $W_{itr}$ is lower than that of $W_{opt}$, set $W_{opt} = W_{itr}$.

Step 6: *Continue iteration with stop criteria*: Go to step 1 for the next iteration, unless the number of iterations exceeds a fixed predetermined value, $I_{max}$. $I_{max}$ depends on network size, the allowable computational time, and the quality of solution desired. Otherwise, the search procedure stops, and $W_{opt}$ is the solution.

In Eq. (8.2), SO considers only the case of $l = 0$, no failure situation, while PSO considers all $l \in F$. In addition, the difference of PSO-W from PSO-L is the set of candidates, $W_{cand}$ in Eq. (8.3). In PSO-L, the candidates are limited to $W_l, l \in F$, while in PSO-W a wider range of candidates is considered.

To clarify the differences between PSO-L and PSO-W, the algorithms are shown in Figures 8.13 and 8.14, respectively. In PSO-L, tabu search is performed for each topology with a single link failure. In PSO-W, all possible single link failures are considered in the process of tabu search.

## 8.4   Performance of PSO-W

The performances of the PSO-W scheme to those of the SO and RO schemes via simulations are compared. The performance measure is the network congestion ratio, $r$. Six sample networks are used, as shown in Figure 8.15, to determine the basic characteristics of these schemes. For the given network topologies, link capacities and traffic demands are randomly generated with

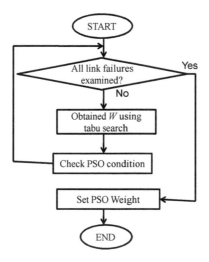

Figure 8.13: Flowchart of PSO-L.

uniform distribution. $C_{max}$ and $I_{max}$ are set to 30 and 1000, respectively, by considering the convergence of the solution carefully.

Let $r(l)$ be denoted as the network congestion ratio for link failure index $l \in F$. To compare $r(l)$ of the different schemes, the network congestion ratios of the PSO-W, SO, and RO schemes with and without a link failure were normalized by that of SO without any link failure. The normalized network congestion ratios are denoted as $r_{PSO-W}(l)$, $r_{SO}(l)$, and $r_{RO}(l)$, respectively.

Table 8.4 compares the worst-case network congestion ratios, $\max_{l \in F} r_{PSO-W}(l)$, $\max_{l \in F} r_{SO}(l)$, and $\max_{l \in F} r_{RO}(l)$, for the sample networks presented in Figure 8.15 for all link-failure scenarios. In terms of the worst-case network congestion ratios over $l$, the following relationship is observed:

$$\max_{l \in F} r_{RO}(l) \leq \max_{l \in F} r_{PSO-W}(l) \leq \max_{l \in F} r_{SO}(l). \tag{8.7}$$

This indicates that PSO-W is able to reduce the worst-case network congestion ratio, unlike SO, while also avoiding the runtime changes of link weights, which would otherwise cause network instability. The reduction ratio of the worst-case congestion ratio, $\alpha$, is defined by

$$\alpha = \frac{\max_{l \in F} r_{SO}(l) - \max_{l \in F} r_{PSO-W}(l)}{\max_{l \in F} r_{SO}(l)}. \tag{8.8}$$

$\alpha$ is also shown in Table 8.4. The range of $\alpha$ is 0.00 to 0.19 for our examined networks.

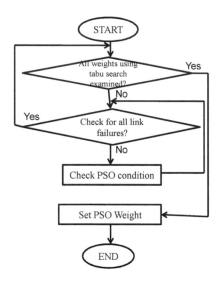

Figure 8.14: Flowchart of PSO-W.

Table 8.5 compares the network congestion ratios with no link failure.

$$r_{SO}(0) = r_{RO}(0) \le r_{PSO-W}(0) \tag{8.9}$$

is observed. When there is no link failure, for the case of $l = 0$, $r_{PSO-W}(0)$ may be higher than $r_{SO}(0)(= r_{RO}(0))$ because the set of link weights of PSO-W is determined so as to reduce the worst-case network congestion ratio. The deviation between $r_{PSO-W}(0)$ and $r_{SO}(0)$, $\beta$, is defined by

$$\beta = \frac{r_{PSO-W}(0) - r_{SO}(0)}{r_{SO}(0)}. \tag{8.10}$$

Table 8.4: Comparison of worst-case network congestion ratios for all link-failure scenarios.

| Network type | $\max_{l \in F} r_{PSO-W}(l)$ | $\max_{l \in F} r_{SO}(l)$ | $\max_{l \in F} r_{RO}(l)$ | $\alpha$ |
|---|---|---|---|---|
| Network 1 | 1.58 | 1.83 | 1.58 | 0.14 |
| Network 2 | 1.36 | 1.67 | 1.27 | 0.19 |
| Network 3 | 1.45 | 1.82 | 1.39 | 0.20 |
| Network 4 | 2.00 | 2.00 | 2.00 | 0.00 |
| Network 5 | 6.95 | 7.04 | 6.59 | 0.06 |
| Network 6 | 1.76 | 1.83 | 1.42 | 0.04 |

(©2011 IEICE, Ref. [15], P. 1970.)

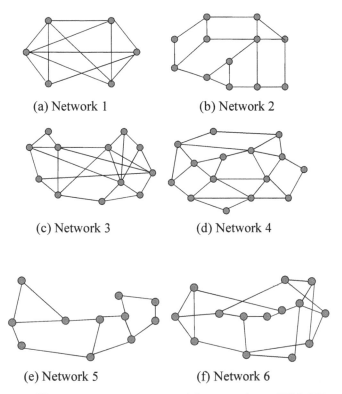

(a) Network 1                            (b) Network 2

(c) Network 3                            (d) Network 4

(e) Network 5                            (f) Network 6

Figure 8.15: Network models to evaluate PSO-W.

$\beta$ is also shown in Table 8.5. To reduce the worst-case network congestion ratio, PSO-W has to pay the penalty of $\beta$ for the case of no link failure.

The performances of PSO-W and PSO-L with respect to the reduction ratio of the worst-case congestion ratio, $\alpha$, are compared. Table 8.6 shows the reduction ratios of PSO-L and PSO-W, $\alpha_{PSO-L}$ and $\alpha_{PSO-W}$, for the sample networks presented in Figure 8.15 for all link-failure scenarios. $\Delta$ is defined as the difference between $\alpha_{PSO-W}$ and $\alpha_{PSO-L}$, which is the reduction ratio of the two different algorithms:

$$\Delta = \alpha_{PSO-W} - \alpha_{PSO-L}. \tag{8.11}$$

We observe that PSO-W finds the more appropriate link weight set than PSO-L, in terms of worst-case performance, in most cases.

Table 8.5: Comparison of network congestion ratios with no link failure.

| Network type | $r_{PSO-W}(0)$ | $r_{SO}(0)(=r_{RO}(0))$ | $\beta$ |
|---|---|---|---|
| Network 1 | 1.08 | 1.00 | 0.08 |
| Network 2 | 1.05 | 1.00 | 0.05 |
| Network 3 | 1.07 | 1.00 | 0.07 |
| Network 4 | 1.00 | 1.00 | 0.00 |
| Network 5 | 4.01 | 1.00 | 3.01 |
| Network 6 | 1.04 | 1.00 | 0.04 |

(©2011 IEICE, Ref. [15], P. 1970.)

Table 8.6: Comparison of network congestion ratios of PSO-W and PSO-L for worst-case.

| Network type | $\alpha_{PSO-W}$ | $\alpha_{PSO-L}$ | $\Delta$ |
|---|---|---|---|
| Network 1 | 0.136 | 0.033 | 0.103 |
| Network 2 | 0.185 | 0.067 | 0.118 |
| Network 3 | 0.202 | 0.164 | 0.038 |
| Network 4 | 0.000 | 0.000 | 0.000 |
| Network 5 | 0.064 | 0.058 | 0.006 |
| Network 6 | 0.038 | 0.027 | 0.011 |

(©2011 IEICE, Ref. [15], P. 1971.)

# Bibliography

[1] J. Moy, "OSPF Version 2," RFC 2328, Apr. 1998.

[2] D. Oran, "OSI IS-IS Intra-domain Routing Protocol," RFC 1142, Feb. 1990.

[3] C. Hedrick, "Routing Information Protocol," RFC 1058, June 1988.

[4] G. Malkin, "Routing Information Protocol," RFC 2453, Nov. 1998.

[5] Y. Rekhter, T. Li, and S. Hares, "A Border Gateway Protocol 4 (BGP-4)," RFC 4271, Jan. 2006.

[6] B. Fortz and M. Thorup, "Optimizing OSPF/IS-IS weights in a changing world," *IEEE Journal on Selected Areas in Communications*, vol. 20, no. 4, pp. 756–767, 2002.

[7] B. Fortz, J. Rexford, and M. Thorup, "Traffic engineering with traditional IP protocols," *IEEE Commun. Mag.*, vol. 40, no. 10, pp. 118–124, 2002.

[8] L.S. Buriol, M.G.C. Resende, C.C. Ribeiro, and M. Thorup, "A hybrid genetic algorithm for the weight setting problem in OSPF-IS-IS routing," *Networks*, vol. 46, no. 1, pp. 36–56, Aug. 2005.

[9] C. Reichert and T. Magedanz, "A fast heuristic for genetic algorithms in link weight optimization," *Lecture Notes in Computer Science*, vol. 3266, pp.144–153, 2004.

[10] F. Glover and M. Laguna, *Tabu Search*, Amsterdam, Kulwer Academic Pulishers, Dordrecht 1997.

[11] A. Nucci and N. Taft, "IGP link weight assignment for operational Tier-1 backbones," IEEE/ACM Transaction on Networking, vol. 15, no. 4, pp. 789–802, Aug. 2007.

[12] G. Iannaccone, C. Chuah, R. Mortier, S. Bhattacharyya, and C. Diot, "Analysis of link failures in a large IP Backbone," in *Proc. 2nd ACM SIGCOM Internet Measurement Workshop* (IMW0, San Francisco, CA, Nov. 2002.

[13] K. Papagiannaki, R. Cruz, and C. Diot, "Network performance monitoring at small time scales, " in *ACM Internet Measurement Conf.*, Miami, FL, Oct. 2003.

[14] M.K Islam and E. Oki, "PSO: Preventive Start-Time Optimization of OSPF Link Weights to Counter Network Failure," *IEEE Commun. Letters*, vol. 14, no. 6, pp. 581–583, June 2010.

[15] M.K Islam and E. Oki, "Optimization of OSPF Link Weights to Counter Network Failure," *IEICE Trans. Commun.*, vol. E94-B, no. 7, pp. 1964–1972, July 2011.

[16] E. Oki and A. Iwaki, "F-TPR: Fine two-phase IP routing scheme over shortest paths for hose model," *IEEE Commun. Letters*, vol. 13, no. 4, pp. 277–279, Apr. 2009.

# Chapter 9

# Mathematical puzzles

This chapter presents mathematical puzzles that can be tackled by integer linear programming (ILP). They are the Sudoku puzzle, a river crossing puzzle, and a lattice puzzle. The ILP formulations and solutions by GLPK are presented. For the river crossing puzzle, the shortest path approach is also introduced to solve the problem.

## 9.1 Sudoku puzzle

### 9.1.1 Overview

Sudoku is a logic-based, combinatorial number-placement puzzle [1]. First published in the United States in 1979, this puzzle was designed by Howard Garns, an architect from Indiana. In its first publication by Dell Magazines, it was known as *Number Place*. The name "Sudoku" was introduced when the puzzle was published in Japan by Nikoli, a Japanese publisher that specializes in games and, especially, logic puzzles. The word "Sudoku" is a Japanese abbreviation for the phrase, "suji wa dokushin ni kagiru," which means that the digits must be remain single. Although it has numerous variants in form, size, and level of difficulty, the Sudoku puzzle most commonly appears in its $9 \times 9$ matrix form.

The rule of Sudoku is simple: fill in an $n \times n$ matrix, which contains some given entries, so that each row, column, and $m \times m$ submatrix, where $n = m^2$, contain each integer 1 through $n$ exactly once. The number and location of the given entries determine the puzzle's level of difficulty. In a $9 \times 9$ puzzle, each row, column, and $3 \times 3$ submatrix should contain the digits 1 through 9 exactly once. An example of the Sudoku problem and the solution is shown in Figure 9.1.

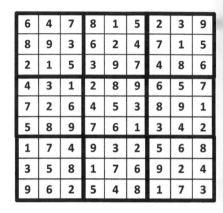

|        | (a) Problem |        |        | (b) Solution |        |

Figure 9.1: Example of $9 \times 9$ Sudoku puzzle.

### 9.1.2 Integer linear programming problem

The Sudoku puzzle is formulated here as an integer linear programming (ILP) task. The Sudoku problem is a satisfiability problem (or feasibility problem) whose goal is to find at least one feasible solution satisfying all of the constraints. In this problem, we do not intend to maximize or minimize any value. Therefore, no objective function must be defined.

For the $9 \times 9$ puzzle, the decision variables, $x_{ijk}$, are defined by

$$x_{ijk} = \begin{cases} 1 & \text{if element } (i,j) \text{ contains integer } k \\ 0 & \text{otherwise,} \end{cases} \qquad (9.1)$$

where $i, j, k \in D$. $D$ is a set of the digits 1 through 9, $D = \{1, \cdots, 9\}$. Let $G$ be a set of $(i, j, k)$, where digit $k$ is given for element $(i, j)$, or $x_{ijk} = 1$.

The $9 \times 9$ puzzle is formulated as the following ILP problem:

$$\text{Constraints} \qquad x_{ijk} = 1, \quad \forall i, j, k \in G \qquad (9.2a)$$

$$\sum_{k=1}^{9} x_{ijk} = 1, \quad \forall i, j \in D \qquad (9.2b)$$

$$\sum_{i=1}^{9} x_{ijk} = 1, \quad \forall j, k \in D \qquad (9.2c)$$

$$\sum_{j=1}^{9} x_{ijk} = 1, \quad \forall i, j \in D \qquad (9.2d)$$

$$\sum_{i=I}^{I+2} \sum_{j=J}^{J+2} x_{ijk} = 1, \quad \forall k \in D, \ I = 1, 4, 7, \ J = 1, 4, 7 \qquad (9.2e)$$

$$x_{ijk} \in \{0, 1\}, \quad \forall i, j, k \in D \qquad (9.2f)$$

Eq. (9.2a) gives the condition that specifies the Sudoku problem. $x_{ijk} = 1$ for $(i, j, k) \in G$ is set. Eq. (9.2b) represents that element $(i, j)$ has only one digit from 1 to 9. Eq. (9.2c) represents that digit $k \in D$ appears once in each column. Eq. (9.2d) represents that digit $k \in D$ appears once in each row. Eq. (9.2e) represents that digit $k \in D$ appears once in each $3 \times 3$ submatrix.

### 9.1.2.1   GLPK listing

Listing 9.1 shows the model file for the Sudoku problem presented in Eqs. (9.2a)–(9.2f). The problem presented in Figure 9.1(a) is written in an input file, as shown in Listing 9.2. After running the command 'glpsol', GLPK reports the solution, as shown in Listing 9.3. Lines 36–71 in Listing 9.1 are the code that outputs the solution in matrix form on the console and the output file.

Listing 9.1: Model file: sudoku9x9.mod

```
1  /* sudoku9x9.mod */
2
3  /* Decision variable */
4
5          var x{i in 1..9, j in 1..9, k in 1..9}, binary;
6          /* x[i,j,k] = 1 means cell[i,j] is assigned number k */
7
8
9  /* Initialization */
10
11         param input_problem{1..9, 1..9}, integer, >=0, <=9, default 0;
12         /* input problem */
13
14         s.t. pre_defined{i in 1..9,j in 1..9,
15                 k in 1..9:input_problem[i,j]!=0}:
16                 x[i,j,k] = (if input_problem[i,j] = k then 1 else 0);
17         /* assign pre-defined numbers */
18
19
```

```
20  /* No objective function */
21
22  /*Constraints */
23
24          s.t. constr_fill{i in 1..9, j in 1..9}: sum{k in 1..9} x[i,j,k] = 1;
25          /* constrain #1 : every cell must be filled by exactly one number */
26
27          s.t. constr_row{i in 1..9, k in 1..9}: sum{j in 1..9} x[i,j,k] = 1;
28          /* constrain #2 : only one k in each row */
29
30          s.t. constr_col{j in 1..9, k in 1..9}: sum{i in 1..9} x[i,j,k] = 1;
31          /* constrain #3 : only one k in each column */
32
33          s.t. constr_sub{I in 1..9 by 3, J in 1..9 by 3, k in 1..9}:
34                  sum{i in I..I+2, j in J..J+2} x[i,j,k] = 1;
35          /* constrain #4 : only one k in each submatrix */
36
37  solve;
38
39
40  printf "This is the solution of sudoku 9x9.\n";
41
42
43  for {i in 1..9}
44  {       for {0..0: i = 1 or i = 4 or i = 7}
45                  printf " +-------+-------+-------+ \n";
46          for {j in 1..9}
47          {       for {0..0: j = 1 or j=4 or j =7}
48                  printf (" |");
49                  printf " %d", sum{k in 1..9} x[i,j,k] * k;
50                  for {0..0: j=9} printf (" |\n");
51          }
52          for {0..0: i = 9}
53                  printf " +-------+-------+-------+ \n";
54  }
55
56  param TXT, symbolic, := "sudoku9x9.txt";
57
58  printf "This is the solution of sudoku 9x9.\n" > TXT;
59
60
61  for {i in 1..9}
62  {       for {0..0: i = 1 or i = 4 or i = 7}
63                  printf " +-------+-------+-------+ \n" >> TXT;
64          for {j in 1..9}
65          {       for {0..0: j = 1 or j=4 or j =7}
66                  printf (" |") >> TXT;
67                  printf " %d", sum{k in 1..9} x[i,j,k] * k >> TXT;
68                  for {0..0: j=9} printf (" |\n") >> TXT;
69          }
70          for {0..0: i = 9}
71                  printf " +-------+-------+-------+ \n" >> TXT;
72  }
```

Listing 9.2: Input file: sudoku9x9.dat

```
1   /* FILE NAME : sudoku9x9.dat */
2   /* This is the Sudoku 9x9 problem */
3
4   data;
5
6   param input_problem : 1 2 3 4 5 6 7 8 9 :=
7                         1 . . 7 8 . 5 2 . .
8                         2 8 . . 6 . 4 . . 5
9                         3 . 1 . . 9 . . 8 .
10                        4 4 . . 2 8 9 . . 7
```

```
                   5 . . . . . . . .
                   6 5 . . 7 6 1 . . 2
                   7 . 7 . . 3 . . 6 .
                   8 3 . . 1 . 6 . . 4
                   9 . . 2 5 . 8 1 . . ;
end;
```

Listing 9.3: Output file: sudoku9x9.txt

```
This is the solution of sudoku 9x9.
+-------+-------+-------+
| 6 4 7 | 8 1 5 | 2 3 9 |
| 8 9 3 | 6 2 4 | 7 1 5 |
| 2 1 5 | 3 9 7 | 4 8 6 |
+-------+-------+-------+
| 4 3 1 | 2 8 9 | 6 5 7 |
| 7 2 6 | 4 5 3 | 8 9 1 |
| 5 8 9 | 7 6 1 | 3 4 2 |
+-------+-------+-------+
| 1 7 4 | 9 3 2 | 5 6 8 |
| 3 5 8 | 1 7 6 | 9 2 4 |
| 9 6 2 | 5 4 8 | 1 7 3 |
+-------+-------+-------+
```

## 9.2   River crossing puzzle

### 9.2.1   Overview

The river crossing puzzle is a logic-based puzzle whose objective is to carry items from one river bank to the other, subject to several constraints. This puzzle can also be classified as a variant of the transport puzzles that include labyrinths, mazes, and sliding puzzles. The level of difficulty of the river crossing puzzle is based on the restrictions on which or how many items can be transported at the same time, or which or how many items may be safely left together. The well-known river crossing puzzles include the problem of the ridge and torch, the problem with the dogs and chicks, the problem of the jealous husbands, and the problem of the wolf, goat, and cabbage. The rules of these problems are described as follows.

**Bridge and torch problem** Four people, A, B, C, and D, come to a river at night. There is a narrow bridge, but it can only hold two people at a time. Because it's night, the torch must be used when people cross the bridge. Unfortunately, the torch can only be used for 15 minutes. Person A can cross the bridge in 1 minute, B in 2 minutes, C in 5 minutes, and D in 8 minutes. How do they all get across the bridge in 15 minutes or less? When two people cross the bridge together, they must move at the slower person's pace [2].

**Dogs and chicks problem** Three dogs and three chicks must cross a river using a boat that can hold at most two items. However, for both banks, if there are chicks present on each bank, they cannot be outnumbered by dogs. How do all the dogs and chicks reach the other bank of the river?

**Jealous husbands problem** Three married couples must cross a river using a boat that can hold at most two people. However, no woman can be in the presence of another man unless her husband is also present. How do all three married couples reach the other bank of the river [3]? This problem is similar to the dogs and the chicks problem. Under the constraint that no woman can be in the presence of another man unless her husband is also present, there is no way the women can outnumber the men on the same bank because they cannot be husbandless.

**Wolf, goat, and cabbage problem** A farmer must transport a wolf, a goat, and a cabbage from one bank of a river to the other using a boat. The boat can only hold a maximum of one item in addition to the farmer. The goat cannot be left alone with the wolf without the farmer's presence, and the cabbage cannot be left alone with the goat without the farmer's presence. How does the farmer bring all items to the other bank of the river [4]?

The following subsections explain how to solve the wolf, goat, and cabbage problem, which is famous as Alcuin's transportation model. This problem is one of the four transportation problems in the book *Propositiones ad Acuandos Iuvenes*, which is attributed to Alcuin of York, an Anglo-Saxon monk and an English leading scholar of his time. Written at the end of the eighth century A.D., the *Propositiones* seems to be the oldest collection of mathematical problems written in Latin [4]. Two different approaches are introduced to solve the problem. They are the integer linear programming (ILP) approach and the shortest path approach.

## 9.2.2   Integer linear programming approach

### 9.2.2.1   Formulation

Similar to the Sudoku puzzle in Section 9.1, the river crossing problem can be solved with the ILP approach. The main idea is how to define the decision variables, the objective function, and the constraints from the problem description. In this puzzle, we have to determine how the farmer can transport all items to the other bank with the minimum number of trips. The ILP approach to solve the wolf, goat, and cabbage problem was presented in [4]. We introduce the key idea of the approach below.

Three vectors, each of which has three components, are defined as the decision variables to represent the configuration of the wolf, the goat, and the cabbage in three different spots. The three vectors $\mathbf{x}$, $\mathbf{y}$, and $\mathbf{z}$ correspond to the configurations of the left bank, the boat, and the right bank, respectively. The first, second, and third components of each vector represent the existences of the wolf, the goat, and the cabbage, respectively.

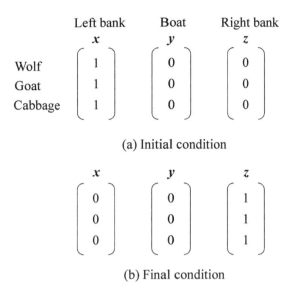

(a) Initial condition

$$
\begin{pmatrix} 0 \\ 0 \\ 0 \end{pmatrix}
\quad
\begin{pmatrix} 0 \\ 0 \\ 0 \end{pmatrix}
\quad
\begin{pmatrix} 1 \\ 1 \\ 1 \end{pmatrix}
$$

$$x \qquad\qquad y \qquad\qquad z$$

(b) Final condition

Figure 9.2: Configurations of **x**, **y**, and **z**.

Figure 9.2(a) represents the initial configuration of the wolf, the goat, and the cabbage. As the initial condition, all three items exist at the left bank, and there is no item on the boat and at the right bank. All the components of **x** are set to 1, and all the components of **y** and **z** are set to 0. Every item that does exist is represented by 1, while every item that does not exist is represented by 0. Figure 9.2(b) represents the final configurations of the wolf, the goat, and the cabbage. As all three items exist at the right bank and there is no item on the boat and at the left bank, all components of **z** are set to 1, and all components of **x** and **y** are set to 0.

Let $x(t, i)$, $y(t, i)$, and $z(t, i)$ be the $i$th components of $\mathbf{x}(t)$, $\mathbf{y}(t)$, and $\mathbf{z}(t)$, respectively, at time $t$. They are defined in the following:

$$
x(t, i) \;=\; \begin{cases} 1 & \text{if item } i \text{ exists at the left bank at time } t, \\ 0 & \text{otherwise.} \end{cases} \tag{9.3a}
$$

$$
y(t, i) \;=\; \begin{cases} 1 & \text{if item } i \text{ exists on the boat at time } t, \\ 0 & \text{otherwise.} \end{cases} \tag{9.3b}
$$

$$
z(t, i) \;=\; \begin{cases} 1 & \text{if item } i \text{ exists at the right bank at time } t, \\ 0 & \text{otherwise.} \end{cases} \tag{9.3c}
$$

$$\tag{9.3d}$$

$t$ indicates the time, which is an integer. At the beginning, $t$ is set to 0. If the boat crosses the river from one bank to the other bank, $t$ is incremented by one. Therefore, if the boat is at the left bank, $t$ is an even number, while, if the boat is at the right bank, $t$ is an odd number. Item $i$ indicates the index for items, where the wolf, the goat, and the cabbage are specified by $i = 1, 2, 3$, respectively. For example, $x(1,2)$ means that the goat exists at the right bank at time $t = 1$.

The wolf, goat, and cabbage problem is formulated below as an ILP problem:

$$\text{Objective}\quad \min\quad w = \sum_{t=0}^{T}\sum_{i=1}^{3} f(t)x(t,i) \tag{9.4a}$$

$$\text{Constraints}\quad \begin{aligned} \mathbf{x}(0) &= (1,1,1) \\ \mathbf{y}(0) &= (0,0,0) \\ \mathbf{z}(0) &= (0,0,0) \end{aligned} \tag{9.4b}$$

$$\left.\begin{aligned} \mathbf{x}(t+1) &= \mathbf{x}(t) - \mathbf{y}(t+1) \\ \mathbf{z}(t+1) &= \mathbf{z}(t) + \mathbf{y}(t+1) \end{aligned}\right\}\quad \forall \text{ even } t \in \Gamma^{-} \tag{9.4c}$$

$$\left.\begin{aligned} \mathbf{x}(t+1) &= \mathbf{x}(t) + \mathbf{y}(t+1) \\ \mathbf{z}(t+1) &= \mathbf{z}(t) - \mathbf{y}(t+1) \end{aligned}\right\}\quad \forall \text{ odd } t \in \Gamma^{-} \tag{9.4d}$$

$$y(t,1) + y(t,2) + y(t,3) \le 1 \quad \forall\ t \in \Gamma \tag{9.4e}$$

$$\left.\begin{aligned} x(t,1) + x(t,2) &\le 1 \\ x(t,2) + x(t,3) &\le 1 \end{aligned}\right\}\quad \forall \text{ odd } t \in \Gamma \tag{9.4f}$$

$$\left.\begin{aligned} -z(t,1) + z(t,2) + z(t,3) &\le 1 \\ z(t,1) + z(t,2) - z(t,3) &\le 1 \end{aligned}\right\}\quad \forall \text{ even } t \in \Gamma \tag{9.4g}$$

$$\mathbf{z}(T) = (1,1,1) \tag{9.4h}$$

$$\mathbf{x}(t), \mathbf{y}(t), \mathbf{z}(t) \in \{0,1\}^{3} \quad \forall\ t \in \Gamma, \tag{9.4i}$$

where $\Gamma = \{0, 1, \cdots, T\}$ and $\Gamma^{-} = \{0, 1, \cdots, T-1\}$. $T$ must be large enough that the final configuration can be obtained.

The objective of the wolf, goat, and cabbage problem is to minimize the number of trips under the condition that the farmer gets all items transported to the other bank without violating the rule. To achieve our goal, the objective function of this problem is represented by Eq. (9.4a). $f(t)$ is the given function that minimizes the number of trips by minimizing the objective function in Eq. (9.4a). How to choose a suitable $f(t)$ is discussed at the end of this subsection.

Constraints in Eqs. (9.4b)–(9.4i) are explained in the following. Eq. (9.4b) represents the initial configuration at $t = 0$, as shown in Figure 9.2(a). Eqs. (9.4c) and (9.4d) represent the state transitions. $\mathbf{x}(t)$ and $\mathbf{z}(t)$ correspond to the wolf-goat-cabbage configuration for the left and right banks, respectively, after the $t$th shipment has been completed, while $\mathbf{y}(t)$ records the boat configuration for the $t$th shipment. Eq. (9.4c) represents the transition when an item is transported from the right bank to the left bank, or when $t$ is even, while Eq. (9.4d) represents the transportation from the left bank to the right bank, or when $t$ is odd. Examples of the state transition are shown in Figure 9.3 for even $t$ and Figure 9.4 for odd $t$. Eq. (9.4e) is introduced because, when the farmer is rowing the boat, he can take at most one item at a time.

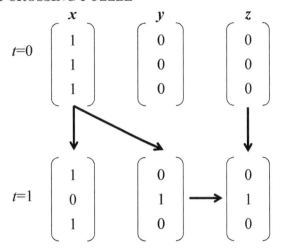

Figure 9.3: Example of state transition for $t$ even.

According to the rule, there are several wolf-goat-cabbage configurations that are not allowed. "Wolf-Goat" and "Goat-Cabbage" are not allowed to be at the same bank without the farmer's presence. Eq. (9.4f) represents the state restriction for odd $t$. It is obtained from the set of possible configurations of $\mathbf{x}(t)$ at the left bank, which is given by

$$\mathbf{x}(t) \in \left\{ \begin{pmatrix} 0 \\ 0 \\ 0 \end{pmatrix}, \begin{pmatrix} 1 \\ 0 \\ 0 \end{pmatrix}, \begin{pmatrix} 0 \\ 1 \\ 0 \end{pmatrix}, \begin{pmatrix} 0 \\ 0 \\ 1 \end{pmatrix}, \begin{pmatrix} 1 \\ 0 \\ 1 \end{pmatrix} \right\} \quad \forall \text{ odd } t \leq \Gamma. \quad (9.5)$$

Eq. (9.4g) represents the state restriction for even $t$. It is obtained from the set of possible configurations of $\mathbf{z}(t)$ at the right bank, which is given by

$$\mathbf{z}(t) \in \left\{ \begin{pmatrix} 0 \\ 0 \\ 0 \end{pmatrix}, \begin{pmatrix} 1 \\ 0 \\ 0 \end{pmatrix}, \begin{pmatrix} 0 \\ 1 \\ 0 \end{pmatrix}, \begin{pmatrix} 0 \\ 0 \\ 1 \end{pmatrix}, \begin{pmatrix} 1 \\ 0 \\ 1 \end{pmatrix}, \begin{pmatrix} 1 \\ 1 \\ 1 \end{pmatrix} \right\} \quad \forall \text{ even } t \leq \Gamma. \quad (9.6)$$

Eq. (9.4h) represents the final configuration, where all the items must be at the right bank, as shown in Figure 9.2(b). Eq. (9.4i) indicates that all the components of $\mathbf{x}(t)$, $\mathbf{y}(t)$, and $\mathbf{z}(t)$ take either 0 or 1.

To solve the ILP problem presented in Eqs. (9.4a)-(9.4i), we must determine the finite time horizon $T$, and $f(t)$ in the objective function.

First, consider how to determine $T$. Let the number of possible states at the left bank be $L$. In the worst case, $L$ states at the left bank are experienced. If there is a feasible solution for the ILP problem, the last state of the left bank should be $\mathbf{x}(t) = (0, 0, 0)$, where $t = 2L - 1$, which is the maximum odd number to be considered. As Eq. (9.5) indicates, $L = 5$, $T = 2 \cdot 5 - 1 = 9$ is determined.

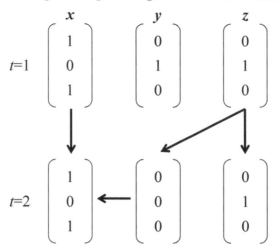

Figure 9.4: Example of state transition for $t$ odd.

Next, consider how to determine a suitable $f(t)$. Intuitively, we find that one of the suitable functions of $f(t)$ is extremely steep with $t$. If $t$ increases, $w$ in Eq. (9.4a) rapidly increases. This means that the farmer must remove all the items from the left bank as soon as possible. In other words, minimizing this objective function is equivalent to minimizing the number of trips. Let us set $f(t)$ as follows:

$$f(t) = I \sum_{t'=0}^{t-1} f(t') + 1, \text{ where } f(0) = 0$$

$$= \begin{cases} 0, & t = 0 \\ (I+1)^{t-1}, & t \geq 1. \end{cases} \tag{9.7}$$

For $I = 3$, $f(t)$ is $f(0) = 0, f(1) = 1, f(2) = 4, f(3) = 16, \cdots$. We are able to prove that $f(t)$ in Eq. (9.7) gives the equivalence between minimizing the objective function in Eq. (9.4a) and minimizing the number of trips using Theorem 9.2.1 as follows.

Assume that all the items are first transported to the right bank by the farmer at $t = t_{vacant}$. This means $x(t, i) = 0$ for $i = 1, 2, 3$ and $t_{vacant} \leq t \leq T$ as no trip is needed after $t = t_{vacant}$. Therefore, the value of the objective function is not increased with $t \geq t_{vacant} - 1$. In addition, at least one item is left at the left bank until $t = t_{vacant} - 1$. In other words, at least $x(t, i) = 1$ for $i$ exists until $t = t_{vacant} - 1$. We would like to minimize $t = t_{vacant}$ to meet our goal. Let one of feasible solutions be $\mathbf{s}$ and the set of feasible solutions be $\mathbf{S}$. Let the value of the objective function with solution $\mathbf{s}$ until time $t$ be denoted as $w(t, \mathbf{s})$.

**Theorem 9.2.1** $w(t, \mathbf{s_1}) > w(t-1, \mathbf{s_2})$ *is satisfied for any feasible solution* $\mathbf{s_1}, \mathbf{s_2} \in \mathbf{S}$, *with* $0 \leq t \leq t_{vacant} - 1$, *if Eq. (9.7) is adopted as* $f(t)$.

**Proof:** Consider $0 \leq t \leq t_{vacant} - 1$. By using this assumption and Eq. (9.7), the lower bound of $w(t, \mathbf{s})$, $\inf\limits_{\mathbf{s} \in \mathbf{S}} w(t, \mathbf{s})$, and the upper bound of $w(t-1, \mathbf{s})$, $\sup\limits_{\mathbf{s} \in \mathbf{S}} w(t-1, \mathbf{s})$, are obtained as follows:

$$
\begin{aligned}
w(t, \mathbf{s}) &= \sum_{t'=0}^{t} \sum_{i=1}^{I} f(t') x(t', i) \\
&= \sum_{t'=0}^{t-1} \sum_{i=1}^{I} f(t') x(t', i) + \sum_{i=1}^{I} f(t) x(t, i) \\
&\geq f(t) \\
&= I \sum_{t'=0}^{t} f(t') + 1 \\
&= \inf_{\mathbf{s_1} \in \mathbf{S}} w(t, \mathbf{s_1}) \quad (9.8)
\end{aligned}
$$

The inequality in Eq. (9.8) is derived as at least $x(t, i) = 1$ for $i$ exists until $t = t_{vacant} - 1$.

$$
\begin{aligned}
w(t-1, \mathbf{s}) &= \sum_{t'=0}^{t-1} \sum_{i=1}^{I} f(t') x(t', i) \\
&\leq I \sum_{t'=0}^{t-1} f(t') \\
&= \sup_{\mathbf{s_2} \in \mathbf{S}} w(t-1, \mathbf{s_2}) \quad (9.9)
\end{aligned}
$$

The inequality in Eq. (9.9) is derived by assuming that all the items are left time until $t - 1$, or $x(t', i) = 1$. Using Eqs. (9.8) and (9.9), we obtain the following relationship:

$$
\inf_{\mathbf{s_1} \in \mathbf{S}} w(t, \mathbf{s_1}) > \sup_{\mathbf{s_2} \in \mathbf{S}} w(t-1, \mathbf{s_2}), \quad \forall \mathbf{s_1}, \mathbf{s_2} \in \mathbf{S}, \quad 0 \leq t \leq t_{vacant} - 1. \quad (9.10)
$$

Therefore, we get

$$
w(t, \mathbf{s_1}) > w(t-1, \mathbf{s_2}), \quad \forall \mathbf{s_1}, \mathbf{s_2} \in \mathbf{S}, \quad 0 \leq t \leq t_{vacant} - 1. \quad (9.11)
$$

∎

Eq. (9.11) indicates that, for any different feasible solution, $\mathbf{s_1}$ and $\mathbf{s_2}$, $w(t, \mathbf{s_1})$ is larger than $w(t-1, \mathbf{s_2})$ with $0 \leq t \leq t_{vacant} - 1$. $w(t, \mathbf{s})$ keeps the same value with $t_{vacant} - 1 \leq t \leq T$, as there are no items at the left bank. Therefore, minimizing $w(t, \mathbf{s})$ for $0 \leq t \leq T$ is equivalent to minimizing the number of trips.

## 9.2.2.2  GLPK listing

Listing 9.4 shows the model file of the wolf, goat, and cabbage problem puzzle presented in Eqs. (9.4a)–(9.4i), where Eq. (9.7) is adopted for $f(t)$. Note that the contribution of $t = 0$ to the objective function is omitted from Listing 9.4, as it is a constant value. After running the command '`glpsol`', GLPK reports the solution, as shown in Listing 9.5, which illustrates the flow of the trips. Lines 74–120 in Listing 9.4 are the code that outputs the illustrations on the console and the output file.

Listing 9.4: Model file: river.mod

```
1   /* river.mod */
2
3   /* Decision variables */
4
5   var x{t in 0..9, i in 1..3}, binary;
6   /* x[t,1] = 1 --> the WOLF is at the left bank, at time t */
7   /* x[t,2] = 1 --> the GOAT is at the left bank, at time t */
8   /* x[t,3] = 1 --> the CABBAGE is at the left bank, at time t */
9
10  var y{t in 0..9, i in 1..3}, binary;
11  /* y[t,1] = 1 --> the WOLF is crossing, at time t */
12  /* y[t,2] = 1 --> the GOAT is crossing, at time t */
13  /* y[t,3] = 1 --> the CABBAGE is crossing, at time t */
14
15  var z{t in 0..9, i in 1..3}, binary;
16  /* z[t,1] = 1 --> the WOLF is at the right bank, at time t */
17  /* z[t,2] = 1 --> the GOAT is at the right bank, at time t */
18  /* z[t,3] = 1 --> the CABBAGE is at the right bank, at time t */
19
20  /* Objective function */
21
22  minimize F: (sum{t in 1..9, i in 1..3} 4^(t-1)*x[t,i]);
23
24  /* Constraints */
25
26          /*Initialization*/
27
28          s.t. init_left{i in 1..3}: x[0,i]=1;
29          /* at t=0, the WOLF, GOAT, and CABBAGE must exist at the left */
30          /* bank */
31
32          s.t. init_cross{i in 1..3}: y[0,i]=0;
33          /* at t=0, nothing is crosing the river */
34
35          s.t. init_right{i in 1..3}: z[0,i]=0;
36          /* at t=0, nothing exists at the right bank */
37
38          /*Transition*/
39
40          s.t. trans_x{t in 0..8, i in 1..3}:
41          x[t+1,i]=(if (t mod 2 == 0)then x[t,i]-y[t+1,i]
42                  else x[t,i]+y[t+1,i]);
43
44          s.t. trans_z{t in 0..8, i in 1..3}:
45          z[t+1,i]=(if (t mod 2 == 0)then z[t,i]+y[t+1,i]
46                  else z[t,i]-y[t+1,i]);
47
48          /*Crossing*/
49
50          s.t. crossing{t in 0..9}: sum{i in 1..3} y[t,i] <= 1;
51          /* When crossing, the farmer brings 1 item or nothing at all */
52
```

```
        /* Posible configuration*/

        s.t. state_odd1{t in 0..9: t mod 2 != 0}: x[t,1]+x[t,2] <= 1;
        /* At the left bank */
        /* The GOAT can not be left with the WOLF without the farmer */

        s.t. state_odd2{t in 0..9: t mod 2 != 0}: x[t,2]+x[t,3] <= 1;
        /* At the left bank */
        /* The CABBAGE can not be left with the GOAT without the farmer */

        s.t. state_even1{t in 0..9: t mod 2 == 0}: -z[t,1]+z[t,2]+z[t,3]<=1;
        /* At right bank */
        /* The CABBAGE can not be left with the GOAT without the farmer */

        s.t. state_even2{t in 0..9: t mod 2 == 0}: z[t,1]+z[t,2]-z[t,3]<=1;
        /* At right bank */
        /* The GOAT can not be left with the WOLF without the farmer */


        /*Final*/

        s.t. final_state{i in 1..3}: z[9,i]=1;
        /* at t =9, the WOLF, GOAT, and CABBAGE must exist at right bank */
solve;

printf "|LEFT_BANK| |CROSSING| |RIGHT_BANK| \n";

for {t in 0..9}
{
        printf " \n";
        printf " \n";
        printf "t = %d", t;
        printf " \n";

        printf: (if x[t,1] = 1 then "wolf\t" else "----\t");
        printf: (if y[t,1] = 1 then "wolf\t" else "----\t");
        printf: (if z[t,1] = 1 then "wolf \n" else "---- \n");

        printf: (if x[t,2] = 1 then "goat\t" else "----\t");
        printf: (if y[t,2] = 1 then "goat\t" else "----\t");
        printf: (if z[t,2] = 1 then "goat \n" else "---- \n");

        printf: (if x[t,3] = 1 then "cabbage\t" else "----\t");
        printf: (if y[t,3] = 1 then "cabbage\t" else "----\t");
        printf: (if z[t,3] = 1 then "cabbage \n" else "---- \n");
}

param TXT, symbolic, := "river.txt";

printf "|LEFT_BANK| |CROSSING| |RIGHT_BANK| \n" > TXT;

for {t in 0..9}
{
        printf " \n" >> TXT;
        printf " \n" >> TXT;
        printf "t = %d", t >> TXT;
        printf " \n" >> TXT;

        printf: (if x[t,1] = 1 then "wolf\t" else "----\t") >> TXT;
        printf: (if y[t,1] = 1 then "wolf\t" else "----\t") >> TXT;
        printf: (if z[t,1] = 1 then "wolf \n" else "---- \n") >> TXT;

        printf: (if x[t,2] = 1 then "goat\t" else "----\t") >> TXT;
        printf: (if y[t,2] = 1 then "goat\t" else "----\t") >> TXT;
        printf: (if z[t,2] = 1 then "goat \n" else "---- \n") >> TXT;
```

```
120    |              printf: (if x[t,3] = 1 then "cabbage\t" else "----\t") >> TXT;
121    |              printf: (if y[t,3] = 1 then "cabbage\t" else "----\t") >> TXT;
122    |              printf: (if z[t,3] = 1 then "cabbage \n" else "---- \n") >> TXT;
123    |  }
```

Listing 9.5: Output file: river.txt

```
1    |LEFT_BANK|  |CROSSING|  |RIGHT_BANK|
2
3
4    t = 0
5    wolf       ----      ----
6    goat       ----      ----
7    cabbage ----         ----
8
9
10   t = 1
11   wolf       ----      ----
12   ----       goat      goat
13   cabbage ----         ----
14
15
16   t = 2
17   wolf       ----      ----
18   ----       ----      goat
19   cabbage ----         ----
20
21
22   t = 3
23   ----       wolf      wolf
24   ----       ----      goat
25   cabbage ----         ----
26
27
28   t = 4
29   ----       ----      wolf
30   goat       goat      ----
31   cabbage ----         ----
32
33
34   t = 5
35   ----       ----      wolf
36   goat       ----      ----
37   ----       cabbage cabbage
38
39
40   t = 6
41   ----       ----      wolf
42   goat       ----      ----
43   ----       ----      cabbage
44
45
46   t = 7
47   ----       ----      wolf
48   ----       goat      goat
49   ----       ----      cabbage
50
51
52   t = 8
53   ----       ----      wolf
54   ----       ----      goat
55   ----       ----      cabbage
56
57
58   t = 9
59   ----       ----      wolf
```

| ---- | ---- | goat    |
|------|------|---------|
| ---- | ---- | cabbage |

As shown in Listing 9.5, at $t = 1$, the farmer takes the goat to the right bank and leaves the wolf and the cabbage at the left bank. At $t = 2$, having put the goat at the right bank, the farmer returns to the left bank. At $t = 3$, the farmer takes the wolf to the right bank. At $t = 4$, having put the wolf at the right bank, the farmer takes the goat and returns to the left bank. At $t = 5$, having left the goat behind at the left bank, the farmer takes the cabbage across to the right bank. At $t = 6$, having put the wolf and the cabbage together at the right bank, the farmer returns to the left bank. At $t = 7$, the farmer picks up the goat and goes across to the right bank. Therefore, at this point, nothing exists at the left bank because all items have been transported to the right bank. Therefore, we can conclude that the minimum number of trips is 7.

### 9.2.3 Shortest path approach

#### 9.2.3.1 Overview

The river crossing puzzle can be solved using the shortest path routing approach. This approach finds the path with minimum cost from the source node to the destination node for the constructed network that represents the feasible states and the transitions for the river crossing puzzle. In the network, each feasible state is denoted as a node, and each transition from one state to another state, or each trip from one bank to the other bank, is denoted as a link with unit cost. Note that the network is an undirected graph, where traffic flows in both directions on each link. After the network is constructed, we can easily solve the problem by determining the shortest path from the source node, which represents the initial state, to the destination node, which represents the final state.

#### 9.2.3.2 Network construction and shortest path

To solve the problem, three steps are considered. The first step is to list the feasible states. The second step is to construct the network. The third step is to find the shortest path.

First, let us list the feasible states. Each state that represents a configuration at the left bank is denoted by $(f, w, g, c)$. $f$, $w$, $g$, and $c$ indicates the existence of the farmer, the wolf, the goat, and the cabbage at the left bank, respectively. If an item exists, the value is set to 1, and otherwise 0. If the state of the left bank is $(f, w, g, c)$, that of the right bank is $(\bar{f}, \bar{w}, \bar{g}, \bar{c})$. If all the states of $(f, w, g, c)$ are counted, there are $16(= 2^4)$ states. However, to satisfy the rule, six states, which are $(0, 0, 1, 1)$, $(1, 1, 0, 0)$, $(0, 1, 1, 0)$, $(1, 0, 0, 1)$, $(0, 1, 1, 1)$, $(1, 0, 0, 0)$, are not allowed and must be removed. Therefore, there are $10 (= 16-6)$ feasible states, as shown in Table 9.1; $(1, 1, 1, 1)$ and $(0, 0, 0, 0)$

Table 9.1: Feasible states at left bank

| Farmer | Wolf | Goat | Cabbage | |
|--------|------|------|---------|---|
| 1 | 1 | 1 | 1 | initial state |
| 1 | 1 | 1 | 0 | |
| 1 | 1 | 0 | 1 | |
| 1 | 0 | 1 | 1 | |
| 1 | 0 | 1 | 0 | |
| 0 | 1 | 0 | 1 | |
| 0 | 1 | 0 | 0 | |
| 0 | 0 | 1 | 0 | |
| 0 | 0 | 0 | 1 | |
| 0 | 0 | 0 | 0 | final state |

indicate the initial and final states, respectively. Changing a state to another state is equivalent to moving from one bank to the other bank.

Second, let us construct the network that represents the feasible states and the state transitions for the wolf, goat, and cabbage problem. Figure 9.5 shows the constructed network. Each node corresponds to each feasible state presented in Table 9.1. Each link indicates a corresponding feasible state transition. Each link cost is set to 1, as the cost of every trip from one bank to the other bank is considered the same. For example, consider the initial state $(1, 1, 1, 1)$. As the farmer is able to cross the river taking the goat from the left side to the right side, $(1, 1, 1, 1)$ is able to move to $(0, 1, 0, 1)$, but is unable to move to any other state. The opposite transition from $(0, 1, 0, 1)$ to $(1, 1, 1, 1)$ is also feasible. Thus, a bidirectional link is set between $(1, 1, 1, 1)$ and $(0, 1, 0, 1)$.

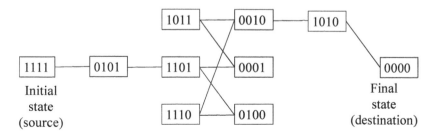

Figure 9.5: Constructed network with unit link costs.

The third step is to find the shortest path from the source node to the destination node. The way to find the shortest path is described in Section 4.1. For the network in Figure 9.5, we find two shortest paths, which are $(0,0,0,0) \rightarrow (0,1,0,1) \rightarrow (1,1,0,1) \rightarrow (0,0,0,1) \rightarrow (1,0,1,1) \rightarrow (0,0,1,0) \rightarrow (1,0,1,0) \rightarrow (0,0,0,0)$ and $(0,0,0,0) \rightarrow (0,1,0,1) \rightarrow (1,1,0,1) \rightarrow (0,1,0,0) \rightarrow (1,1,1,0) \rightarrow (0,0,1,0) \rightarrow (1,0,1,0) \rightarrow (0,0,0,0)$. Both paths give the minimum number of trips, which is 7.

The interpretation of the first path is the same as the solution described in Section 9.2.2. We describe the interpretation of the second path is as follows. At the first trip, the farmer takes the goat to the right bank and leaves the wolf and the cabbage at the left bank. On the second trip, having put the goat at the right bank, the farmer returns to the left bank. On the third trip, the farmer takes the cabbage to the right bank. On the fourth trip, having put the cabbage at the right bank, the farmer takes the goat and returns to the left bank. On the fifth trip, having left the goat behind at the left bank, the farmer takes the wolf across to the right bank. On the sixth trip, having put the wolf and the cabbage together at the right bank, the farmer returns to the left bank. On the seventh trip, the farmer takes the goat across to the right bank. Therefore, at this point, there is nothing at the left bank because all items have been transported to the right bank. The first solution and the second solution are different in what the farmer picks up—the wolf or the cabbage on the third trip. Both solutions are acceptable.

### 9.2.4 Comparison of two approaches

It is presented that the wolf, goat, and cabbage problem can be solved by both the ILP and shortest path approaches in Sections 9.2.2 and 9.2.3, respectively. Both approaches find the solution in the minimum number of trips and also the flow of the trips.

The two approaches have their own benefits and challenges. The ILP approach does not always need to list all feasible states and their transitions, but it counts on being able to define a mathematical model from the rules of the problem. In addition, it is not always possible to define the problem in linear form. In some cases, the river crossing problem is formulated as a non-linear integer problem, or a combinatorial problem. On the other hand, the shortest path approach counts on being able to construct a network by listing all possible states and their transitions. However, once the connected network is constructed, it is guaranteed that the problem is solved by the approach.

## 9.3 Lattice puzzle

### 9.3.1 Overview

Similar to the Sudoku puzzle, the lattice puzzle is a logic-based, number-placement puzzle. There are several types of lattice, either in two dimensions

(2D) or three dimensions (3D). A 2D rectangular lattice is introduced here.

The lattice puzzle has a simpler rule than the Sudoku puzzle: fill in every cell of the $n \times n$ lattice with a positive integer number ($\geq 0$), so that the sum of numbers in every $m \times l$ and $l \times m$ sublattice equals $k$. Figure 9.6 shows an example of the lattice puzzle with $n = 6$, $m = 3$, $l = 2$, and $k = 7$. There is no number assigned to the lattice in the initial condition.

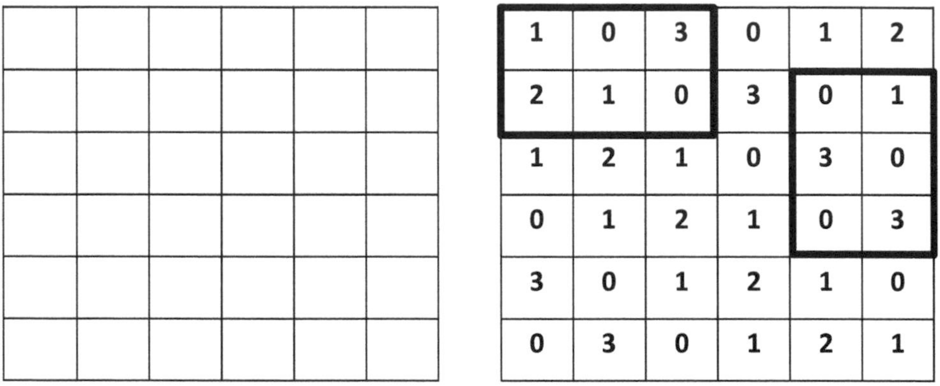

(a) Problem                    (b) Solution

Figure 9.6: $n \times n$ lattice puzzle. The sum of numbers in every $m \times l$ and $l \times m$ sublattice must be $k$, where $n = 6$, $m = 3$, $l = 2$, and $k = 7$.

## 9.3.2   Integer linear programming

The problem of the lattice puzzle is formulated as integer linear programming (ILP). The lattice puzzle is also a satisfiability problem (or feasibility problem) whose goal is to find at least one feasible solution satisfying all the constraints. Once again, no objective function must be defined, because we do not intend to maximize or minimize any value.

Let us consider the lattice puzzle with $n = 6$, $m = 3$, $l = 2$, and $k = 7$. The decision variables, $x_{ij} \geq 0$, are denoted, where $i \in N$ and $j \in N$, and $N = \{1, 2, \cdots, 6\}$. The problem of the lattice puzzle is formulated as the following ILP problem:

$$\text{Constraints} \quad \sum_{i=I}^{I+2} \sum_{j=J}^{J+1} x_{ij} = 7, \forall \, 1 \leq I \leq 4, 1 \leq J \leq 5 \qquad (9.12a)$$

$$\sum_{i=I}^{I+1} \sum_{j=J}^{J+2} x_{ij} = 7, \forall \, 1 \leq I \leq 5, 1 \leq J \leq 4, \qquad (9.12b)$$

where $I$ and $J$ are integers. Eq. (9.12a) represents that the sum of numbers in every $3 \times 2$ sublattice must be 7. Eq. (9.12b) represents that the sum of numbers in every $2 \times 3$ sublattice must be 7.

### 9.3.2.1 GLPK listing

Listing 9.6 shows the model file for the lattice puzzle problem presented in Eqs. (9.12a) and (9.12b). After running the command 'glpsol', GLPK reports the solution, as shown in Listing 9.7. Lines 23-58 in Listing 9.1 are the code that outputs the solution in lattice form on the console and output file.

Listing 9.6: Model file: lattice6x6.mod

```
/* lattice6x6.mod */

/* Decision variable */

        var x{i in 1..6, j in 1..6}, integer, >= 0;
        /* x[i,j] = 1 means cell[i,j] is assigned by number 1 */
        /* x[i,j] = 7 means cell[i,j] is assigned by number 7 */


/* No objective function */


/* Constraints */

s.t. constr_3x2{I in 1..4, J in 1..5}:
                sum{i in I..I+2, j in J..J+1} x[i,j] = 7;
        /* constrain #1: The sum of the numbers in any 3x2 sublattice is 7
*/

s.t. constr_2x3{I in 1..5, J in 1..4}:
                sum{i in I..I+1, j in J..J+2} x[i,j] = 7;
        /* constrain #2: The sum of the numbers in any 2x3 sublattice is 7
*/

solve;

printf "This is the solution of lattice 6x6. \n";


for {i in 1..6}
{
        printf " +----+----+----+----+----+----+\n";
        for {j in 1..6}
        {   printf(" |");
                printf " %d ", x[i,j];
                for {0..0: j=6} printf(" |\n");
        }
        for {0..0: i=6}
        {       printf " +----+----+----+----+----+----+\n";
        }
}

param TXT, symbolic, := "lattice6x6.txt";

printf "This is the solution of lattice 6x6. \n" > TXT;


for {i in 1..6}
{
```

```
49 |          printf " +----+----+----+----+----+----+\n" >> TXT;
50 |          for {j in 1..6}
51 |          {    printf(" |") >> TXT;
52 |                  printf " %d ", x[i,j] >> TXT;
53 |                  for {0..0: j=6} printf(" |\n") >> TXT;
54 |          }
55 |          for {0..0: i=6}
56 |          {        printf " +----+----+----+----+----+----+\n" >> TXT;
57 |          }
58 | }
```

Listing 9.7: Output file: lattice6x6.txt

```
 1 | This is the solution of lattice 6x6.
 2 | +----+----+----+----+----+----+
 3 | | 1  | 0  | 3  | 0  | 1  | 2  |
 4 | +----+----+----+----+----+----+
 5 | | 2  | 1  | 0  | 3  | 0  | 1  |
 6 | +----+----+----+----+----+----+
 7 | | 1  | 2  | 1  | 0  | 3  | 0  |
 8 | +----+----+----+----+----+----+
 9 | | 0  | 1  | 2  | 1  | 0  | 3  |
10 | +----+----+----+----+----+----+
11 | | 3  | 0  | 1  | 2  | 1  | 0  |
12 | +----+----+----+----+----+----+
13 | | 0  | 3  | 0  | 1  | 2  | 1  |
14 | +----+----+----+----+----+----+
```

# Bibliography

[1] A.C. Bartlett and A.N. Langville, "An integer programming model for the Sudoku problem," 2006. Available at http://langvillea.people.cofc.edu/sudoku.

[2] G. Gribakin, "Some simple and not so simple maths problem," Gleb Gribakin's website, retrieved from http://web.am.qub.ac.uk/users/g.gribakin/problems.html, 2012.

[3] I. Pressman and D. Singmaster, "The jealous husbands and the missionaries and cannibals," *The Mathematical Gazette*, vol 73, no. 464, pp. 73–81, 1989.

[4] R. Borndörfer, Martin, Grötschel, and Andreas Löbel, "Alcuin's transportation problems and integer programming," Konrad-Zuse-Zentrum für Informationstechnik Berlin, Preprint SC-95-27, 1995.

## Exercise 9.1

Solve the Sudoku puzzles, as shown in Figure 9.7.

## Exercise 9.2

Determine the mathematical model for the $16 \times 16$ Sudoku puzzle, as shown in Figure 9.8, and solve it.

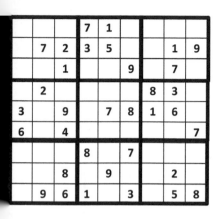

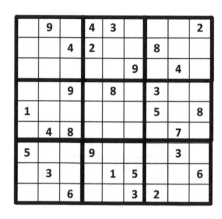

(a) Problem 1                    (b) Problem 2

Figure 9.7: $9 \times 9$ Sudoku puzzles.

## Exercise 9.3

Solve the dogs and chicks problem. Three dogs and three chicks must cross a river using a boat that can hold at most two items. However, for both banks, if there are chicks present on each bank, they cannot be outnumbered by dogs. How do all the dogs and chicks reach the other bank of the river? Check if each approach presented in Sections 9.2.2 and 9.2.3 is applicable to solve the problem. If not, explain the reason.

## Exercise 9.4

Solve the jealous husbands problem. Three married couples must cross a river using a boat that can hold at most two people. However, no woman can be in the presence of another man unless her husband is also present. How do all three married couples reach the other bank of the river? Check if each approach presented in Sections 9.2.2 and 9.2.3 is applicable to solve the problem. If not, explain the reason.

## Exercise 9.5

Solve the lattice puzzle with $n = 6$, $m = 3$, $l = 2$, and $k = 13$. Fill in every cell of $n \times n$ lattice with a positive integer number ($\geq 0$), so that the sum of numbers in every $m \times l$ and $l \times m$ sublattice can be equal to $k$.

| | 15 | | 13 | 12 | | 10 | | 8 | 7 | 6 | | | 3 | | 1 |
|---|---|---|---|---|---|---|---|---|---|---|---|---|---|---|---|
| 12 | | | 9 | | 15 | | 13 | | 3 | 2 | 1 | 8 | 7 | 6 | 5 |
| 8 | | | 5 | 4 | 3 | | 1 | | 15 | | 13 | | 11 | | 9 |
| | | 2 | 1 | | | 6 | 5 | | 11 | | 9 | | 15 | | 13 |
| 15 | | | 14 | | | 9 | | | 8 | | 6 | | 4 | | 2 |
| | | 9 | | 15 | 16 | 13 | | 3 | 4 | 1 | | 7 | | | 6 |
| | 8 | 5 | | 3 | | | 2 | | | 13 | | 11 | 12 | 9 | 10 |
| 3 | | | | 7 | 8 | | 6 | | 12 | | 10 | | 16 | | |
| | | 16 | 15 | | 9 | 12 | | 6 | | 8 | | 2 | 1 | | 3 |
| | 9 | | | 14 | | 16 | | 2 | | 4 | 3 | | | 8 | |
| 6 | | | 7 | | 1 | | 3 | | 13 | | | 10 | | 12 | |
| | | 4 | | 6 | | 8 | | 10 | | 12 | | 14 | | 16 | 15 |
| | 14 | | 16 | | 10 | 11 | 12 | | | 7 | 8 | | | | |
| 9 | 10 | 11 | | 13 | | 15 | 16 | | | 3 | 4 | | 6 | | 8 |
| 5 | 6 | 7 | 8 | | | | 4 | | 14 | | 16 | | | 11 | 12 |
| | | | 4 | | 6 | 7 | 8 | 9 | | 11 | 12 | 13 | | 15 | 16 |

Figure 9.8: $16 \times 16$ Sudoku puzzle.

# Appendix A

# Derivation of Eqs. (7.6a)–(7.6c) for hose model

Eqs. (7.5a)–(7.5c), which is the LP problem of finding $\boldsymbol{T_H} = \{t_{pq}\}$ that maximizes link load on $(i, j)$, is represented with a matrix expression by

$$\max \boldsymbol{X}_{ij}^T \boldsymbol{t} \tag{A.1a}$$

$$s.t. \quad \boldsymbol{At} \leq \boldsymbol{C} \tag{A.1b}$$

$$\boldsymbol{t} \geq 0, \tag{A.1c}$$

where

$$\boldsymbol{t}^T = [t_{11}t_{12}\cdots t_{1N}|\cdots|t_{N1}t_{N2}\cdots t_{NN}] \tag{A.2a}$$

$$\boldsymbol{X}_{ij}^T = [x_{ij}^{11}x_{ij}^{12}\cdots x_{ij}^{1N}|\cdots|x_{ij}^{N1}x_{ij}^{N2}\cdots x_{ij}^{NN}] \tag{A.2b}$$

$$\boldsymbol{A} = \begin{bmatrix}
1 & 1 & \cdots & 1 & 0 & 0 & \cdots & 0 & 0 & 0 & \cdots & 0 & \cdots & 0 & 0 & \cdots & 0 \\
0 & 0 & \cdots & 0 & 1 & 1 & \cdots & 1 & 0 & 0 & \cdots & 0 & \cdots & 0 & 0 & \cdots & 0 \\
0 & 0 & \cdots & 0 & 0 & 0 & \cdots & 0 & 1 & 1 & \cdots & 1 & \cdots & 0 & 0 & \cdots & 0 \\
 & & \cdots & & & & \cdots & & & & \cdots & & \cdots & & & \cdots & \\
0 & 0 & \cdots & 0 & 0 & 0 & \cdots & 0 & 0 & 0 & \cdots & 0 & \cdots & 1 & 1 & \cdots & 1 \\
1 & 0 & \cdots & 0 & 1 & 0 & \cdots & 0 & 1 & 0 & \cdots & 0 & \cdots & 1 & 0 & \cdots & 0 \\
0 & 1 & \cdots & 0 & 0 & 1 & \cdots & 0 & 0 & 1 & \cdots & 0 & \cdots & 0 & 1 & \cdots & 0 \\
 & & \cdots & & & & \cdots & & & & \cdots & & \cdots & & & \cdots & \\
0 & 0 & \cdots & 1 & 0 & 0 & \cdots & 1 & 0 & 0 & \cdots & 1 & \cdots & 0 & 0 & \cdots & 1
\end{bmatrix} \tag{A.2c}$$

$$\boldsymbol{C}^T = [\alpha_1\alpha_2\cdots\alpha_N|\beta_1\beta_2\cdots\beta_N] \tag{A.2d}$$

$N$ is the number of nodes. $\boldsymbol{t}$ is an $NN \times 1$ matrix. $\boldsymbol{X}_{ij}$ is an $NN \times 1$ matrix. $\boldsymbol{A}$ is a $2N \times NN$ matrix. $\boldsymbol{C}$ is a $2N \times 1$ matrix.

The dual of the LP problem represented by Eqs. (A.1a)–(A.2d) for $(i,j)$ is,

$$\min \boldsymbol{C}^T \boldsymbol{z}_{ij} \tag{A.3a}$$

$$s.t. \quad \boldsymbol{A}^T \boldsymbol{z} \geq \boldsymbol{X}_{ij} \tag{A.3b}$$

$$\boldsymbol{z}_{ij} \geq 0, \tag{A.3c}$$

where

$$\boldsymbol{z}_{ij}^T \;=\; [\pi_{ij}(1)\pi_{ij}(2)\cdots\pi_{ij}(N)|\lambda_{ij}(1)\lambda_{ij}(2)\cdots\lambda_{ij}(N)]. \tag{A.4a}$$

$\boldsymbol{z}_{ij}$ is a $2N \times 1$ matrix. Eqs. (A.3a) and (A.4a) and Eqs. (A.2b)–(A.2d) is a matrix expression of Eqs. (7.6a)–(7.6c).

# Appendix B

# Derivation of Eqs. (7.12a)–(7.12c) for HSDT model

Eqs. (7.11a)–(7.11d), which is the LP problem of finding $\boldsymbol{T} = \{t_{pq}\}$ that maximizes link load on $(i, j)$, is represented with a matrix expression by

$$\max \boldsymbol{X}_{ij}^T \boldsymbol{t} \tag{B.1a}$$

$$s.t. \quad \boldsymbol{At} \leq \boldsymbol{C} \tag{B.1b}$$

$$\boldsymbol{t} \geq 0, \tag{B.1c}$$

where

$$\boldsymbol{t}^T = [t_{11}t_{12}\cdots t_{1N}|\cdots|t_{N1}t_{N2}\cdots t_{NN}] \tag{B.2a}$$

$$\boldsymbol{X}_{ij}^T = [x_{ij}^{11}x_{ij}^{12}\cdots x_{ij}^{1N}|\cdots|x_{ij}^{N1}x_{ij}^{N2}\cdots x_{ij}^{NN}] \tag{B.2b}$$

$$
\boldsymbol{A} =
\left[
\begin{array}{cccc|cccc|cccc|c|cccc}
1 & 1 & \cdots & 1 & 0 & 0 & \cdots & 0 & 0 & 0 & \cdots & 0 & \cdots & 0 & 0 & \cdots & 0 \\
0 & 0 & \cdots & 0 & 1 & 1 & \cdots & 1 & 0 & 0 & \cdots & 0 & \cdots & 0 & 0 & \cdots & 0 \\
0 & 0 & \cdots & 0 & 0 & 0 & \cdots & 0 & 1 & 1 & \cdots & 1 & \cdots & 0 & 0 & \cdots & 0 \\
 & & \cdots & & & & \cdots & & & & \cdots & & & & & \cdots & \\
0 & 0 & \cdots & 0 & 0 & 0 & \cdots & 0 & 0 & 0 & \cdots & 0 & \cdots & 1 & 1 & \cdots & 1 \\
\hline
1 & 0 & \cdots & 0 & 1 & 0 & \cdots & 0 & 1 & 0 & \cdots & 0 & \cdots & 1 & 0 & \cdots & 0 \\
0 & 1 & \cdots & 0 & 0 & 1 & \cdots & 0 & 0 & 1 & \cdots & 0 & \cdots & 0 & 1 & \cdots & 0 \\
 & & \cdots & & & & \cdots & & & & \cdots & & & & & \cdots & \\
0 & 0 & \cdots & 1 & 0 & 0 & \cdots & 1 & 0 & 0 & \cdots & 1 & \cdots & 0 & 0 & \cdots & 1 \\
\hline
1 & 0 & \cdots & 0 & 0 & 0 & \cdots & 0 & 0 & 0 & \cdots & 0 & \cdots & 0 & 0 & \cdots & 0 \\
0 & 1 & \cdots & 0 & 0 & 0 & \cdots & 0 & 0 & 0 & \cdots & 0 & \cdots & 0 & 0 & \cdots & 0 \\
 & & \cdots & & & & \cdots & & & & \cdots & & & & & \cdots & \\
0 & 0 & \cdots & 1 & 0 & 0 & \cdots & 0 & 0 & 0 & \cdots & 0 & \cdots & 0 & 0 & \cdots & 0 \\
\hline
 & & \cdots & & & & \cdots & & & & \cdots & & \cdots & & & \cdots & \\
\hline
0 & 0 & \cdots & 0 & 0 & 0 & \cdots & 0 & 0 & 0 & \cdots & 0 & \cdots & 1 & 0 & \cdots & 0 \\
0 & 0 & \cdots & 0 & 0 & 0 & \cdots & 0 & 0 & 0 & \cdots & 0 & \cdots & 0 & 1 & \cdots & 0 \\
 & & \cdots & & & & \cdots & & & & \cdots & & & & & \cdots & \\
0 & 0 & \cdots & 0 & 0 & 0 & \cdots & 0 & 0 & 0 & \cdots & 0 & \cdots & 0 & 0 & \cdots & 1 \\
\hline
-1 & 0 & \cdots & 0 & 0 & 0 & \cdots & 0 & 0 & 0 & \cdots & 0 & \cdots & 0 & 0 & \cdots & 0 \\
0 & -1 & \cdots & 0 & 0 & 0 & \cdots & 0 & 0 & 0 & \cdots & 0 & \cdots & 0 & 0 & \cdots & 0 \\
 & & \cdots & & & & \cdots & & & & \cdots & & & & & \cdots & \\
0 & 0 & \cdots & -1 & 0 & 0 & \cdots & 0 & 0 & 0 & \cdots & 0 & \cdots & 0 & 0 & \cdots & 0 \\
\hline
 & & \cdots & & & & \cdots & & & & \cdots & & \cdots & & & \cdots & \\
\hline
0 & 0 & \cdots & 0 & 0 & 0 & \cdots & 0 & 0 & 0 & \cdots & 0 & \cdots & -1 & 0 & \cdots & 0 \\
0 & 0 & \cdots & 0 & 0 & 0 & \cdots & 0 & 0 & 0 & \cdots & 0 & \cdots & 0 & -1 & \cdots & 0 \\
 & & \cdots & & & & \cdots & & & & \cdots & & & & & \cdots & \\
0 & 0 & \cdots & 0 & 0 & 0 & \cdots & 0 & 0 & 0 & \cdots & 0 & \cdots & 0 & 0 & \cdots & -1 \\
\end{array}
\right]
\tag{B.2c}
$$

$$
\begin{aligned}
\boldsymbol{C} = [&\alpha_1\alpha_2\cdots\alpha_N|\beta_1\beta_2\cdots\beta_N| \\
&\gamma_{11}\gamma_{12}\cdots\gamma_{1N}|\cdots|\gamma_{N1}\gamma_{N2}\cdots\gamma_{NN}| \\
&-\delta_{11}-\delta_{12}\cdots-\delta_{1N}|\cdots|-\delta_{N1}-\delta_{N2}\cdots-\delta_{NN}].
\end{aligned}
\tag{B.2d}
$$

$N$ is the number of nodes. $\boldsymbol{t}$ is an $NN \times 1$ matrix. $\boldsymbol{X}_{ij}$ is an $NN \times 1$ matrix. $\boldsymbol{A}$ is a $(2N + 2NN) \times NN$ matrix. $\boldsymbol{C}$ is a $(2N + 2NN) \times 1$ matrix.

The dual of the LP problem represented by Eqs. (B.1a)–(B.2d) for $(i, j)$ is

$$\min \boldsymbol{C}^T \boldsymbol{z}_{ij} \tag{B.3a}$$

$$s.t. \quad \boldsymbol{A}^T \boldsymbol{z} \geq \boldsymbol{X}_{ij} \tag{B.3b}$$

$$\boldsymbol{z}_{ij} \geq 0, \tag{B.3c}$$

where

$$
\begin{aligned}
z_{ij}^T = \; & [\pi_{ij}(1)\pi_{ij}(2)\cdots\pi_{ij}(N)|\lambda_{ij}(1)\lambda_{ij}(2)\cdots\lambda_{ij}(N)| \\
& \eta_{ij}(1,1)\eta_{ij}(1,2)\cdots\eta_{ij}(1,N)|\cdots|\eta_{ij}(N,1)\eta_{ij}(N,2)\cdots\eta_{ij}(N,N)| \\
& \theta_{ij}(1,1)\theta_{ij}(1,2)\cdots\theta_{ij}(1,N)|\cdots|\theta_{ij}(N,1)\theta_{ij}(N,2)\cdots\theta_{ij}(N,N)].
\end{aligned}
$$

$$(B.4a)$$

$z_{ij}$ is a $(2N + 2NN) \times 1$ matrix. Eqs. (B.3a) and (B.4a) and Eqs. (B.2b)–(B.2d) is a matrix expression of Eqs. (7.12a)–(7.12c).

In the HSDT model, for $A$ and $C$, while the first $2N$ rows, which correspond to Eqs. (7.11b) and (7.11c), are the same as those of the hose model, the next $2NN$ rows, which correspond to Eq. (7.11d), are newly introduced. For $z_{ij}$, while the first $2N$ rows are the same as those of the hose model, the next $2NN$ rows are also newly introduced. As a result, the introduced $2NN$ rows in $z_{ij}$ produce the term of $\sum_{p,q\in Q}[\gamma_{pq}\eta_{ij}(p,q) - \delta_{pq}\theta_{ij}(p,q)]$ in Eq. (7.13d) and that of $\eta_{ij}(p,q) - \theta_{ij}(p,q)$ in Eq. (7.13e).

# Appendix C

# Derivation of Eqs. (7.16a)–(7.16d) for HLT model

Eqs. (7.15a)–(7.15d), which is the LP problem of finding $\boldsymbol{T} = \{t_{pq}\}$ that maximizes link load on $(i, j)$, is represented with a matrix expression by

$$\max \boldsymbol{X}_{ij}^T \boldsymbol{t} \tag{C.1a}$$

$$\text{s.t.} \quad \boldsymbol{A}\boldsymbol{t} \leq \boldsymbol{C} \tag{C.1b}$$

$$\boldsymbol{t} \geq 0, \tag{C.1c}$$

where

$$\boldsymbol{t}^T = [t_{11}t_{12}\cdots t_{1N}|\cdots|t_{N1}t_{N2}\cdots t_{NN}] \qquad (\text{C.2a})$$

$$\boldsymbol{X}_{ij}^T = [x_{ij}^{11}x_{ij}^{12}\cdots x_{ij}^{1N}|\cdots|x_{ij}^{N1}x_{ij}^{N2}\cdots x_{ij}^{NN}] \qquad (\text{C.2b})$$

$$
\boldsymbol{A} = \begin{bmatrix}
1 & 1 & \cdots & 1 & 0 & 0 & \cdots & 0 & \cdots & 0 & 0 & \cdots & 0 \\
0 & 0 & \cdots & 0 & 1 & 1 & \cdots & 1 & \cdots & 0 & 0 & \cdots & 0 \\
& & \cdots & & & & \cdots & & \cdots & & & \cdots & \\
0 & 0 & \cdots & 0 & 0 & 0 & \cdots & 0 & \cdots & 1 & 1 & \cdots & 1 \\
1 & 0 & \cdots & 0 & 1 & 0 & \cdots & 0 & \cdots & 1 & 0 & \cdots & 0 \\
0 & 1 & \cdots & 0 & 0 & 1 & \cdots & 0 & \cdots & 0 & 1 & \cdots & 0 \\
& & \cdots & & & & \cdots & & \cdots & & & \cdots & \\
0 & 0 & \cdots & 1 & 0 & 0 & \cdots & 1 & \cdots & 0 & 0 & \cdots & 1 \\
a_{11}^{11} & a_{12}^{11} & \cdots & a_{1N}^{11} & a_{21}^{11} & a_{22}^{11} & \cdots & a_{2N}^{11} & \cdots & a_{31}^{11} & a_{32}^{11} & \cdots & a_{3N}^{11} \\
a_{11}^{12} & a_{12}^{12} & \cdots & a_{1N}^{12} & a_{21}^{12} & a_{22}^{12} & \cdots & a_{2N}^{12} & \cdots & a_{31}^{12} & a_{32}^{12} & \cdots & a_{3N}^{12} \\
& & \cdots & & & & \cdots & & \cdots & & & \cdots & \\
a_{11}^{1N} & a_{12}^{1N} & \cdots & a_{1N}^{1N} & a_{21}^{1N} & a_{22}^{1N} & \cdots & a_{2N}^{1N} & \cdots & a_{31}^{1N} & a_{32}^{1N} & \cdots & a_{3N}^{1N} \\
& & \cdots & & & & \cdots & & \cdots & & & \cdots & \\
a_{11}^{N1} & a_{12}^{N1} & \cdots & a_{1N}^{N1} & a_{21}^{N1} & a_{22}^{N1} & \cdots & a_{2N}^{N1} & \cdots & a_{31}^{N1} & a_{32}^{N1} & \cdots & a_{3N}^{N1} \\
a_{11}^{N2} & a_{12}^{N2} & \cdots & a_{1N}^{N2} & a_{21}^{N2} & a_{22}^{N2} & \cdots & a_{2N}^{N2} & \cdots & a_{31}^{N2} & a_{32}^{N2} & \cdots & a_{3N}^{N2} \\
& & \cdots & & & & \cdots & & \cdots & & & \cdots & \\
a_{11}^{NN} & a_{12}^{NN} & \cdots & a_{1N}^{NN} & a_{21}^{NN} & a_{22}^{NN} & \cdots & a_{2N}^{NN} & \cdots & a_{31}^{NN} & a_{32}^{NN} & \cdots & a_{3N}^{NN}
\end{bmatrix}
$$

$$(\text{C.2c})$$

$$\boldsymbol{C} = [\alpha_1\alpha_2\cdots\alpha_N|\beta_1\beta_2\cdots\beta_N|y_{11}\cdots y_{1N}|\cdots|y_{N1}\cdots y_{NN}] \qquad (\text{C.2d})$$

$N$ is the number of nodes. $\boldsymbol{t}$ is an $NN \times 1$ matrix. $\boldsymbol{X}_{ij}$ is an $NN \times 1$ matrix. $\boldsymbol{A}$ is a $(2N + NN) \times NN$ matrix. $\boldsymbol{C}$ is a $(2N + NN) \times 1$ matrix.

The dual of the LP problem represented by Eqs. (C.1a)–(C.2d) for $(i, j)$ is

$$\min \boldsymbol{C}^T \boldsymbol{z}_{ij} \qquad (\text{C.3a})$$

$$s.t. \quad \boldsymbol{A}^T \boldsymbol{z} \geq \boldsymbol{X}_{ij} \qquad (\text{C.3b})$$

$$\boldsymbol{z}_{ij} \geq 0, \qquad (\text{C.3c})$$

where

$$
\begin{aligned}
\boldsymbol{z}_{ij}^T =\ & [\pi_{ij}(1)\pi_{ij}(2)\cdots\pi_{ij}(N)|\lambda_{ij}(1)\lambda_{ij}(2)\cdots\lambda_{ij}(N)| \\
& \theta_{ij}(1,1)\theta_{ij}(1,2)\cdots\theta_{ij}(1,N)].
\end{aligned}
$$

$$(\text{C.4a})$$

$\boldsymbol{z}_{ij}$ is a $(2N+NN)\times 1$ matrix. Eqs. (C.3a) and (C.4a) and Eqs. (C.2b)–(C.2d) is a matrix expression of Eqs. (7.16a)–(7.16c).

In the HLT model, for $\boldsymbol{A}$ and $\boldsymbol{C}$, while the first $2N$ rows, which correspond to Eqs. (7.15b) and (7.15c), are the same as those of the hose model, the next $NN$ rows, which correspond to Eq. (7.15d), are newly introduced. For $\boldsymbol{z}_{ij}$, while the first $2N$ rows are the same as those of the hose, the next $NN$ rows are also newly introduced. As a result, the introduced $NN$ rows in $z_{ij}$ produce the term of $\sum_{(s,t)\in E} y_s t \theta_{ij}(s,t)$ in Eq. (7.17d) and that of $\sum_{(s,t)\in E} a_{st}^{pq} \theta_{ij}(s,t)$ in Eq. (7.17e).

# Answers to exercises

**Answer** 2.1

Figure D.1 shows the feasible region and the optimum solution. Let $z$ be the objective function $z = 8x_1 + 6x_2$. We want to maximize objective function $z$. We rewrite this function as $x_2 = -\frac{4}{3} + \frac{z}{6}$. The slope of this function is $-\frac{4}{3}$, and it intersects the $x_2$-axis at $(0, z)$. If we move this function up along the $x_2$-axis while keeping its slope, $z$ increases. On the other hand, if we move it down along the $x_2$-axis, $z$ decreases. As shown in Figure D.1, the maximum value of $z$ is determined by moving the objective function up along the $x_2$-axis while retaining the slope, $-\frac{4}{3}$, under the condition that the function passes through the feasible region. We obtain the maximum value of $z = 132$, when $x_2 = -\frac{4}{3} + \frac{z}{6}$ passes through $(x_1, x_2) = (12, 6)$. We can obtain the same solution by the simplex method.

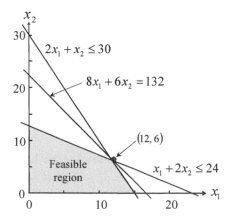

Figure D.1: Feasible region and optimum solution.

**Answer 2.2**

We show the optimum solution by checking every corner point. Figure D.2 shows the feasible region and every corner point. Table D.1 shows all values of the objective function for every corner point. At the corner of $(\frac{12}{5}, \frac{42}{5})$, $10x_1 + 12x_2 = \frac{624}{5} = 124.8$ is the maximum value along the values of the objective function associated with every corner point.

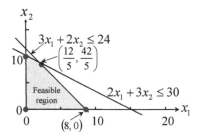

Figure D.2: Feasible region and corner points.

Table D.1: Values of $10x_1 + 12x_2$ in corner points

| Corner point $(x_1, x_2)$ | $10x_1 + 12x_2$ |
|:---:|:---:|
| $(0,0)$ | $0$ |
| $(0,10)$ | $120$ |
| $(\frac{12}{5}, \frac{42}{5})$ | $\frac{624}{5}(=124.8)$ |
| $(8,0)$ | $80$ |

**Answer 2.3**

We show the optimum solution using the simplex method in Figure D.3. There are three corner points, which are $(0, \frac{3}{10})$, $(2, \frac{1}{10})$, and $(5, 0)$. Let us start at the corner point of $(0, \frac{3}{10})$. The value of the objective function at $(0, \frac{3}{10})$ is 360. We move to $(2, \frac{1}{10})$. The value of the objective function at $(2, \frac{1}{10})$ is 280, which is decreased compared with the value of 360 associated with $(0, \frac{3}{10})$. If we continue to move from $(2, \frac{1}{10})$ to $(5, 0)$, the value of the objective function, which is 400, increases. Therefore, the corner point of $(2, \frac{1}{10})$ gives the minimum solution, where the value of the objective function is 280.

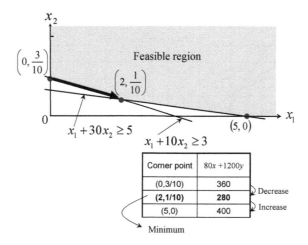

Figure D.3: Solution by simplex method.

**Answer** 2.4

Figure D.4 shows the feasible region and the optimum solution as found by ILP. We find that the optimum solution is $(x_1, x_2) = (0, 10)$, and the maximum value of the objective function is 120.

**Answer** 2.5

Figure D.5 shows the feasible region and the optimum solution in ILP. We find that the optimum solution is $(x_1, x_2) = (4, 6)$, and the maximum value of the objective function is 108.

**Answer** 3.1

1. Let $x_1$, $x_2$, $x_3$, and $x_4$ be the raw materials A, B, C, and D (kg), respectively. Let $z$ be the nutrient cost. The optimization problem of minimiz-

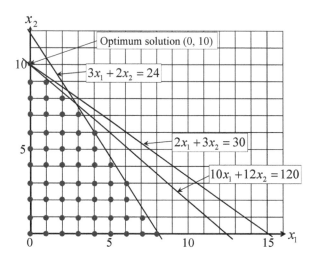

Figure D.4: Feasible region and optimum solution in ILP.

ing $z$ is formulated as an LP problem as follows:

| | | |
|---|---|---|
| Objective | min $z = 5.00x_1 + 7.50x_2 + 3.75x_3 + 2.50x_4$ | (D.5a) |
| Constraints | $0.18x_1 + 0.31x_2 + 0.12x_3 + 0.18x_4 \geq 18$ | (D.5b) |
| | $0.43x_1 + 0.25x_2 + 0.12x_3 + 0.50x_4 \geq 31$ | (D.5c) |
| | $0.31x_1 + 0.37x_2 + 0.37x_3 + 0.12x_4 \geq 25$ | (D.5d) |
| | $x_1 \geq 0$ | (D.5e) |
| | $x_2 \geq 0$ | (D.5f) |
| | $x_3 \geq 0$ | (D.5g) |
| | $x_4 \geq 0$ | (D.5h) |

2. By solving the LP problem in Eqs. (D.5a)–(D.5h), we find that the optimum solution is $(x_1, x_2, x_3, x_4) = (0, 0, 44.83, 70.12)$, and the minimum value of the objective function is $z_{min} \approx 343.39$.

3. The dual problem of the primal problem in Eqs. (D.5a)–(D.5h) is for-

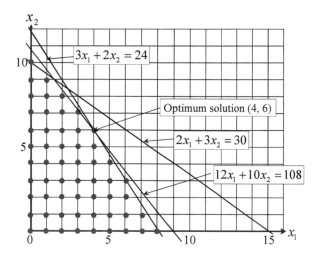

Figure D.5: Feasible region and optimum solution in ILP.

mulated as follows:

$$\text{Objective} \quad \max \quad w = 18y_1 + 31y_2 + 25y_3 \quad \text{(D.6a)}$$

$$\text{Constraints} \quad 0.18y_1 + 0.43y_2 + 0.31y_3 \le 5.00 \quad \text{(D.6b)}$$

$$0.31y_1 + 0.25y_2 + 0.37y_3 \le 7.50 \quad \text{(D.6c)}$$

$$0.12y_1 + 0.12y_2 + 0.37y_3 \le 3.75 \quad \text{(D.6d)}$$

$$0.18y_1 + 0.50y_2 + 0.12y_3 \le 2.50 \quad \text{(D.6e)}$$

$$y_1 \ge 0 \quad \text{(D.6f)}$$

$$y_2 \ge 0 \quad \text{(D.6g)}$$

$$y_3 \ge 0 \quad \text{(D.6h)}$$

4. By solving the dual problem in Eqs. (D.6a)–(D.6h), we find that the optimum solution is $(y_1, y_2, y_3) = (9.10, 0, 7.18)$, and the maximum value of the objective function is $w_{max} \approx 343.39$. We confirm $z_{min} = w_{max}$.

## Answer 3.2

1. Let $x_1$, $x_2$, and $x_3$ be quantities of regular shampoo, exclusive shampoo, and conditioner (liters), respectively. Let $z$ be the profit. The optimiza-

tion problem of maximizing $z$ is formulated as an LP problem as follows:

| | | |
|---|---|---|
| Objective | max $\quad z = 1.5x_1 + 2.0x_2 + 2.5x_3$ | (D.7a) |
| Constraints | $0.3x_1 + 0.5x_2 + 0.2x_3 \leq 100$ | (D.7b) |
| | $0.6x_1 + 0.3x_2 + 0.1x_3 \leq 150$ | (D.7c) |
| | $0.1x_1 + 0.2x_2 + 0.7x_3 \leq 200$ | (D.7d) |
| | $x_1 \geq 0$ | (D.7e) |
| | $x_2 \geq 30$ | (D.7f) |
| | $x_3 \geq 0$ | (D.7g) |

2. By solving the LP problem in Eqs. (D.7a)–(D.7g), we find that the optimum solution is $(x_1, x_2, x_3) = (108.85, 30, 261.579)$, and the maximum value of the objective function is $z_{max} \approx 877.37$.

3. The dual problem of the primal problem in Eqs. (D.7a)–(D.7g) is formulated as follows:

| | | |
|---|---|---|
| Objective | min $\quad w = 100y_1 + 150y_2 + 200y_3 - 30y_4$ | (D.8a) |
| Constraints | $0.3y_1 + 0.6y_2 + 0.1y_3 \geq 1.5$ | (D.8b) |
| | $0.5y_1 + 0.3y_2 + 0.2y_3 - y_4 \geq 2.0$ | (D.8c) |
| | $0.2y_1 + 0.1y_2 + 0.7y_3 \geq 2.5$ | (D.8d) |
| | $y_1 \geq 0$ | (D.8e) |
| | $y_2 \geq 0$ | (D.8f) |
| | $y_3 \geq 0$ | (D.8g) |
| | $y_4 \geq 0$ | (D.8h) |

4. By solving the dual problem in Eqs. (D.8a)–(D.8h), we find that the optimum solution is $(y_1, y_2, y_3, y_4) = (4.21, 0, 2.37, 0.58)$, and the minimum value of the objective function is $w_{min} \approx 877.37$. We confirm $z_{max} = w_{min}$.

**Answer 4.1**

We can use the model file as shown in Listing 4.4 and make an input file to express the network in Figure 4.14. We can also use Dijkstra's algorithm. The shortest path from node 1 to node 6 is $1 \rightarrow 4 \rightarrow 5 \rightarrow 6$, and the cost of the path is 4. The shortest path from node 2 to node 6 is $2 \rightarrow 4 \rightarrow 5 \rightarrow 6$, and the cost of the path is 5.

**Answer 4.2**

We delete link $(4, 5)$ of the network in Figure 4.14, or set the cost of the link to infinity ($\infty$). The shortest path from node 1 to node 6 is $1 \rightarrow 3 \rightarrow 5 \rightarrow 6$, and the cost of the path is 7. The shortest path from node 2 to node 6 is $2 \rightarrow 3 \rightarrow 5 \rightarrow 6$, and the cost of the path is 6.

**Answer 4.3**
We can use the model file as shown in Listing 4.16 and make an input file to express the network in Figure 4.14. We can also use the cycle-canceling algorithm. Figure D.6 shows a solution of the minimum-cost flow problem. The traffic of the volume of $v = 80$ is divided into five paths, from $v_1$ to $v_5$. $v_1 = 30$ is sent on the first path, $1 \rightarrow 3 \rightarrow 6$. $v_2 = 10$ is sent on the second path, $1 \rightarrow 3 \rightarrow 5 \rightarrow 6$. $v_3 = 10$ is sent on the third path, $1 \rightarrow 2 \rightarrow 3 \rightarrow 5 \rightarrow 6$. $v_4 = 10$ is sent on the fourth path, $1 \rightarrow 2 \rightarrow 4 \rightarrow 5 \rightarrow 6$. $v_5 = 10$ is sent on the fifth path, $1 \rightarrow 4 \rightarrow 5 \rightarrow 6$. The total cost is 610.

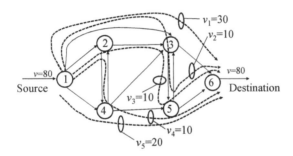

Figure D.6: Solution of minimum-cost flow problem.

**Answer 4.4**
We can use the model file as shown in Listing 4.12, and make an input file to express the network in Figure 4.15. We can also use the Ford-Fulkerson algorithm. Figure D.7 shown a solution of the max flow problem. The maximum traffic volume from node 1 to node 6 is $v = 103$ and consists of five paths with their corresponding traffic volumes of $v_1$ to $v_5$. $v_1 = 13$ is sent on the first path, $1 \rightarrow 2 \rightarrow 5 \rightarrow 6$. $v_2 = 10$ is sent on the second path, $1 \rightarrow 2 \rightarrow 3 \rightarrow 6$. $v_3 = 32$ is sent on the third path, $1 \rightarrow 3 \rightarrow 6$. $v_4 = 35$ is sent on the fourth path, $1 \rightarrow 4 \rightarrow 3 \rightarrow 6$. $v_5 = 13$ is sent on the fifth path, $1 \rightarrow 4 \rightarrow 6$.

**Answer 4.5**
We can use the model file as shown in Listing 4.16 and make an input file

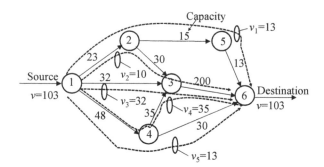

Figure D.7: Solution of max flow problem.

to express the network in Figure 4.16. We can also use the cycle-canceling algorithm. Figure D.8 shows a solution of the minimum-cost flow problem. The traffic of the volume of $v = 20$ is divided into three paths, from $v_1$ to $v_3$. $v_1 = 10$ is sent on the first path, $1 \rightarrow 2 \rightarrow 4$. $v_2 = 20$ is sent on the second path, $1 \rightarrow 2 \rightarrow 3 \rightarrow 4$. $v_3 = 10$ is sent on the third path, $1 \rightarrow 3 \rightarrow 4$. The total cost is 202.

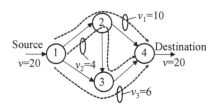

Figure D.8: Solution of minimum-cost flow problem.

**Answer 5.1**
We can use the model file as shown in Listing 5.3 and make an input file to express the wavelength assignment problem in Figure 5.14. We can also use the disjoint shortest pair algorithm or Suurballe's algorithm. We find two disjoint routes of $1 \rightarrow 8 \rightarrow 3 \rightarrow 4 \rightarrow 10 \rightarrow 7 \rightarrow 12$ and $1 \rightarrow 2 \rightarrow 9 \rightarrow 5 \rightarrow 6 \rightarrow 11 \rightarrow 12$. The minimum value, which is the total costs of the two paths, is 18.

**Answer** 6.1

We can use the model file as shown in Listing 6.1 and make an input file to express the wavelength assignment problem in Figure 6.7. Figure D.9 shows a solution of the wavelength assignment. $\lambda 2$ is is assigned to path 1. $\lambda 3$ is assigned to path 2. $\lambda 1$ is assigned to path 3. $\lambda 1$ is assigned to path 4. $\lambda 2$ is assigned to path 5. The minimum required number of wavelengths is 3.

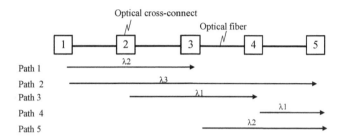

Figure D.9: Solution of wavelength assignment problem.

**Answer** 7.1

$r_P = 0.700$, $r_H = 0.933$, $r_{HSDT1} = 0.800$, $r_{HSDT2} = 0.875$, and $r_{HLT} = 0.700$ are obtained. The relationship of $r_P = r_{HLT} < r_{HSDT1} < r_{HSDT2} < r_H$ is obtained.

**Answer** 7.2

$r_P = 0.583$, $r_H = 0.875$, $r_{HSDT1} = 0.658$, $r_{HSDT2} = 0.760$, and $r_{HLT} = 0.621$ are obtained. The relationship of $r_P < r_{HLT} < r_{HSDT1} < r_{HSDT2} < r_H$ is obtained.

**Answer** 7.3

**Proof of Property 2:**

We extend Proof 1 that was applied to the hose model to prove Property 2. ("only if" direction): Let routing $x_{ij}^{pq}$ have congestion ratio $\leq r$ for all traffic matrices constrained by the intermediate model. (i.e., $\sum_{p,q \in Q} x_{ij}^{pq} t_{pq} \leq c_{ij} \cdot r$ for all $(i,j)$). The problem of finding $\boldsymbol{T} = \{t_{pq}\}$ that maximizes link load on

$(i, j)$ is formulated as the following LP problem:

$$\max \sum_{p,q \in Q} x_{ij}^{pq} t_{pq} \tag{D.9a}$$

$$s.t. \sum_{q \in Q} t_{pq} \le \alpha_p, \quad p \in Q \tag{D.9b}$$

$$\sum_{p \in Q} t_{pq} \le \beta_q, \quad q \in Q \tag{D.9c}$$

$$\delta_{pq} \le t_{pq} \le \gamma_{pq}, \quad p,q \in Q, \tag{D.9d}$$

The decision variables are $t_{pq}$. The given parameters are $x_{ij}^{pq}$, $\alpha_p$, $\beta_q$, $\delta_{pq}$, and $\gamma_{pq}$. The dual of the LP problem in Eqs. (D.9a)–(D.9d) for $(i, j)$ is:

$$\min \sum_{p \in Q} \alpha_p \pi_{ij}(p) + \sum_{p \in Q} \beta_p \lambda_{ij}(p)$$

$$+ \sum_{p,q \in Q} [\gamma_{pq} \eta_{ij}(p, q) - \delta_{pq} \theta_{ij}(p, q)] \tag{D.10a}$$

$$s.t. \quad x_{ij}^{pq} \le \pi_{ij}(p) + \lambda_{ij}(q) + \eta_{ij}(p, q) - \theta_{ij}(p, q),$$

$$\forall p, q \in Q, (i, j) \in E \tag{D.10b}$$

$$\pi_{ij}(p), \lambda_{ij}(p), \eta_{ij}(p, q), \theta_{ij}(p, q) \ge 0, \quad \forall p, q \in Q, (i, j) \in E. \tag{D.10c}$$

The derivation of Eqs. (D.10a)–(D.10c) is described in Appendix B. Because of $\sum_{pq} x_{ij}^{pq} t_{pq} \le c_{ij} \cdot r$ in Eq. (D.9a), the dual, $\sum_{p \in Q} \alpha_p \pi_{ij}(p) + \sum_{p \in Q} \beta_p \lambda_{ij}(p) + \sum_{p,q \in Q} [\gamma_{pq} \eta_{ij}(p, q) - \delta_{pq} \theta_{ij}(p, q)]$ in Eq. (D.10a), for any $(i, j)$, must have the same optimal value. The optimal value in Eq. (D.10a) should be $\le c_{ij} \cdot r$. Therefore, the objective function of the dual satisfies (i). Requirement (ii) is satisfied by dual problem constraint (D.10b).
("if" direction): Let $x_{ij}^{pq}$ be a routing, and $T = \{t_{pq}\}$ be any valid traffic matrix. Let $\pi_{ij}(p)$, $\lambda_{ij}(p)$, $\delta_{pq}$, and $\theta_{ij}(p, q)$ be the parameters satisfying requirements (i) and (ii). Consider $(i, j) \in E$. From (ii) we have

$$x_{ij}^{pq} \le \pi_{ij}(p) + \lambda_{ij}(q) + \eta_{ij}(p, q) - \theta_{ij}(p, q).$$

Summing over all edge node pairs $(p, q)$, we have

$$\sum_{p,q \in Q} x_{ij}^{pq} t_{pq} \le \sum_{p,q \in Q} [\pi_{ij}(p) + \lambda_{ij}(q) + \eta_{ij}(p, q) - \theta_{ij}(p, q)] t_{pq}$$

$$= \sum_{p \in Q} \pi_{ij}(p) \sum_{q \in Q} t_{pq} + \sum_{q \in Q} \lambda_{ij}(q) \sum_{p \in Q} t_{pq}$$

$$+ \sum_{p,q \in Q} [\eta_{ij}(p, q) - \theta_{ij}(p, q)] t_{pq}$$

$$\le \sum_{p \in Q} \pi_{ij}(p) \alpha_p + \sum_{p \in Q} \lambda_{ij}(p) \beta_p$$

$$+ \sum_{p,q \in Q} [\gamma_{pq} \eta_{ij}(p, q) - \delta_{pq} \theta_{ij}(p, q)].$$

The last equality is obtained using the constraints of the intermediate model. From (i), we have

$$\sum_{p,q \in Q} x_{ij}^{pq} t_{pq} \leq \sum_{p \in Q} \pi_{ij}(p)\alpha_p + \sum_{p \in Q} \lambda_{ij}(p)\beta_p$$
$$+ \sum_{p,q \in Q} [\gamma_{pq}\eta_{ij}(p,q) - \delta_{pq}\theta_{ij}(p,q)]$$
$$\leq c_{ij} \cdot r.$$

This indicates that for any traffic matrix constrained by the intermediate model, the load on any link is at most $r$. ∎

**Answer 7.4**

**Proof of Property 3:**

We extend Proof 1 that was applied to the hose model to prove Property 3.

("only if" direction): Let routing $x_{ij}^{pq}$ have congestion ratio $\leq r$ for all traffic matrices constrained by the HLT model. (i.e., $\sum_{p,q \in Q} x^{pq} \leq c_{ij} \cdot r$ for all $(i,j)$) The problem of finding $T = \{t_{pq}\}$ that maximizes link load on $(i,j)$ is formulated as the following LP problem:

$$\max \sum_{p,q \in Q} x_{ij}^{pq} t_{pq} \tag{D.11a}$$

$$s.t. \sum_{q \in Q} d_{pq} \leq \alpha_p, \quad \forall p \in Q \tag{D.11b}$$

$$\sum_{p \in Q} t_{pq} \leq \beta_q, \quad \forall q \in Q \tag{D.11c}$$

$$\sum_{p,q \in Q} a_{ij}^{pq} t_{pq} \leq y_{ij}, \forall, (i,j) \in E \tag{D.11d}$$

The decision variables are $t_{pq}$. The given parameters are $x_{ij}^{pq}$, $\alpha_p$, $\beta_q$, $a_{ij}^{pq}$, and $y_{ij}$. The dual of the LP problem in Eqs. (D.11a)–(D.11d) for $(i,j)$ is

$$\min \sum_{p \in Q} \alpha_p \pi_{ij}(p) + \sum_{p \in Q} \beta_p \lambda_{ij}(p) + \sum_{(s,t) \in E} \theta_{ij}(s,t) y_{st} \tag{D.12a}$$

$$s.t. \quad \pi_{ij}(p) + \lambda_{ij}(q) + \sum_{(s,t) \in E} a_{st}^{pq} \theta_{ij}(s,t) \geq x_{ij}^{pq},$$

$$\forall p, q \in Q, (i,j) \in E \tag{D.12b}$$

$$\pi_{ij}(p), \lambda_{ij}(p) \geq 0, \quad \forall p \in Q, \quad (i,j) \in E \tag{D.12c}$$

$$\theta_{ij}(s,t) \geq 0, \quad \forall (i,j), (s,t) \in E. \tag{D.12d}$$

The derivation of Eqs. (D.12a)–(D.12d) is described in Appendix C. Because of $\sum_{pq} x_{ij}^{pq} t_{pq} \leq c_{ij} \cdot r$ in Eq. (D.11a), the dual, $\sum_{p \in Q} \alpha_p \pi_{ij}(p) + \sum_{p \in Q} \beta_p \lambda_{ij}(p) +$

$\sum_{(s,t)\in E} \theta_{ij}(s,t)y_{st}$ in Eq. (D.12a), for any $(i,j)$, must have the same optimal value. The optimal value in Eq. (D.12a) should be $\leq c_{ij} \cdot r$. Therefore, the objective function of the equivalent satisfies (i). Requirement (ii) is satisfied by equivalent problem constraint (D.12b).

("if" direction): Let $x_{ij}^{pq}$ be a routing, and $T = \{t_{pq}\}$ be any valid traffic matrix. Let $\pi_{ij}(p)$, $\lambda_{ij}(p)$, and $\theta_{ij}(s,t)$ be the parameters satisfying requirements (i) and (ii). Consider $(i,j) \in E$. From (ii) we have

$$x_{ij}^{pq} \leq \pi_{ij}(p) + \lambda_{ij}(q) + \sum_{(s,t)\in E} a_{st}^{pq}\theta_{ij}(s,t).$$

Summing over all edge node pairs $(p,q)$, we have

$$\sum_{p,q\in Q} x_{ij}^{pq}t_{pq} \leq \sum_{p,q\in Q}\Big(\pi_{ij}(p) + \lambda_{ij}(q) + \sum_{(s,t)\in E} a_{st}^{pq}\theta_{ij}(s,t)\Big)t_{pq}$$

$$= \sum_{(s,t)\in E} \theta_{ij}(s,t)\Big(\sum_{p,q\in Q} a_{st}^{pq}t_{pq}\Big)$$

$$+ \sum_{p\in Q}\pi_{ij}(p)\sum_{q\in Q}t_{pq} + \sum_{q\in Q}\lambda_{ij}(q)\sum_{p\in Q}t_{pq}$$

$$\leq \sum_{(s,t)\in E}\theta_{ij}(s,t)y_{st}$$

$$+ \sum_{p\in Q}\pi_{ij}(p)\alpha_p + \sum_{p\in Q}\lambda_{ij}(p)\beta_p.$$

The last equality is obtained using the constraints of the HLT model. From (i), we have

$$\sum_{p,q\in Q} x_{ij}^{pq}t_{pq} \leq \sum_{(s,t)\in E}\theta_{ij}(s,t)y_{st} + \sum_{p\in Q}\alpha_p\pi_{ij}(p) + \sum_{p\in Q}\beta_p\lambda_{ij}(p)$$

$$\leq c_{ij} \cdot r.$$

This indicates that for any traffic matrix constrained by the HLT model, the load on any link is at most $r$. ∎

**Answer 9.1**
Solution for problem (a):

```
+-------+-------+-------+
| 9 6 5 | 7 1 2 | 4 8 3 |
| 8 7 2 | 3 5 4 | 6 1 9 |
| 4 3 1 | 6 8 9 | 2 7 5 |
+-------+-------+-------+
```

```
| 1 2 7 | 9 6 5 | 8 3 4 |
| 3 5 9 | 4 7 8 | 1 6 2 |
| 6 8 4 | 2 3 1 | 5 9 7 |
+-------+-------+-------+
| 5 1 3 | 8 2 7 | 9 4 6 |
| 7 4 8 | 5 9 6 | 3 2 1 |
| 2 9 6 | 1 4 3 | 7 5 8 |
+-------+-------+-------+
```

Solution for problem (b):

```
+-------+-------+-------+
| 7 9 5 | 4 3 8 | 6 1 2 |
| 3 6 4 | 2 7 1 | 8 9 5 |
| 8 2 1 | 5 6 9 | 7 4 3 |
+-------+-------+-------+
| 2 5 9 | 1 8 7 | 3 6 4 |
| 1 7 3 | 6 9 4 | 5 2 8 |
| 6 4 8 | 3 5 2 | 1 7 9 |
+-------+-------+-------+
| 5 8 7 | 9 2 6 | 4 3 1 |
| 4 3 2 | 7 1 5 | 9 8 6 |
| 9 1 6 | 8 4 3 | 2 5 7 |
+-------+-------+-------+
```

**Answer** 9.2

Listing D.1: Model file: sudoku16x16-q2.mod

```
 1  /* sudoku16x16-q2.mod */
 2
 3
 4  /* Decision Variable */
 5
 6          var x{i in 1..16, j in 1..16, k in 1..16}, binary;
 7          /* x[i,j,k] = 1 means cell[i,j] is assigned number k */
 8
 9  /* Initialization */
10
11          param input_problem{1..16, 1..16}, integer, >=0, <=16, default 0;
12          /* input problem */
13
14  s.t. pre_defined{i in 1..16, j in 1..16, k in 1..16: input_problem[i,j]!=0}:
15                      x[i,j,k] = (if input_problem[i,j]= k then 1 else 0);
16          /* assign pre-defined number */
17
```

```
18  /* No Objective Function */
19
20
21  /* Constrains */
22
23          s.t. constr_fill{i in 1..16, j in 1..16}: sum{k in 1..16}x[i,j,k]=1;
24          /* constrain #1 : every position in matrix must be filled */
25
26          s.t. constr_row{i in 1..16, k in 1..16}: sum{j in 1..16} x[i,j,k]=1;
27          /* constrain #2 : only one k in each row */
28
29          s.t. constr_col{j in 1..16, k in 1..16}: sum{i in 1..16} x[i,j,k]=1;
30          /* constrain #3 : only one k in each column */
31
32          s.t. constr_sub{I in 1..16 by 4, J in 1..16 by 4, k in 1..16}:
33                      sum{i in I..I+3, j in J..J+3} x[i,j,k] = 1;
34          /* constrain #4 : only one k in each submatrix */
35
36  solve;
37
38  printf "This is the solution of sudoku 16x16. \n";
39
40
41  for {i in 1..16}
42  {       for {0..0: i = 1 or i = 5 or i = 9 or i = 13}
43  printf " +-------------+-------------+-------------+------------+ \n";
44          for {j in 1..16}
45          {   for {0..0: j = 1 or j = 5 or j = 9 or j = 13}
46                  printf(" |");
47                  printf " %2d", sum{k in 1..16} x[i,j,k] * k;
48                  for {0..0: j=16} printf(" |\n");
49          }
50          for {0..0: i = 16}
51  printf " +-------------+-------------+-------------+------------+ \n";
52  }
53
54  param TXT, symbolic, := "sudoku16x16.txt";
55
56  printf "This is the solution of sudoku 16x16. \n" > TXT;
57
58
59  for {i in 1..16}
60  {       for {0..0: i = 1 or i = 5 or i = 9 or i = 13}
61  printf " +-------------+-------------+-------------+------------+ \n">> TXT;
62          for {j in 1..16}
63          {   for {0..0: j = 1 or j = 5 or j = 9 or j = 13}
64                  printf(" |") >> TXT;
65                  printf " %2d", sum{k in 1..16} x[i,j,k] * k >> TXT;
66                  for {0..0: j=16} printf(" |\n") >> TXT;
67          }
68          for {0..0: i = 16}
69  printf " +-------------+-------------+-------------+------------+ \n">> TXT;
70  }
```

Listing D.2: Input file: sudoku16x16-q2.dat

```
1   /* sudoku16x16-q2.dat */
2
3   data;
4
5   param input_problem :     1  2  3  4  5  6  7  8  9  10 11 12 13 14 15 16:=
6                          1  .  15 .  13 12 .  10 .  8  7  6  .  .  3  .  1
7                          2  12 .  .  9  .  15 .  13 .  3  2  1  8  7  6  5
8                          3  8  .  .  5  4  3  .  1  .  15 .  13 .  11 .  9
9                          4  .  .  2  1  .  .  6  5  .  11 .  9  .  15 .  13
10                         5  15 .  .  14 .  .  9  .  .  8  .  6  .  4  .  2
```

```
                        6  .  .  9  . 15 16 13  .  3  4  1  .  7  .  .  6
                        7  .  8  5  .  3  .  .  2  .  . 13  . 11 12  9 10
                        8  3  .  .  .  7  8  .  6  . 12  . 10  . 16  .  .
                        9  .  . 16 15  .  9 12  .  6  .  8  .  2  1  .  3
                       10  .  9  .  . 14  . 16  .  2  .  4  3  .  .  8  .
                       11  6  .  .  7  .  1  .  3  . 13  .  . 10  . 12  .
                       12  .  .  4  .  6  .  8  . 10  . 12  . 14  . 16 15
                       13  . 14  . 16  . 10 11 12  .  .  7  8  .  .  .  .
                       14  9 10 11  . 13  . 15 16  .  .  3  4  .  6  .  8
                       15  5  6  7  8  .  .  .  4  . 14  . 16  .  . 11 12
                       16  .  .  .  4  .  6  7  8  9  . 11 12 13  . 15 16;
end;
```

Solution for problem

```
+-------------+-------------+-------------+-------------+
| 16 15 14 13 | 12 11 10  9 |  8  7  6  5 |  4  3  2  1 |
| 12 11 10  9 | 16 15 14 13 |  4  3  2  1 |  8  7  6  5 |
|  8  7  6  5 |  4  3  2  1 | 16 15 14 13 | 12 11 10  9 |
|  4  3  2  1 |  8  7  6  5 | 12 11 10  9 | 16 15 14 13 |
+-------------+-------------+-------------+-------------+
| 15 16 13 14 | 11 12  9 10 |  7  8  5  6 |  3  4  1  2 |
| 11 12  9 10 | 15 16 13 14 |  3  4  1  2 |  7  8  5  6 |
|  7  8  5  6 |  3  4  1  2 | 15 16 13 14 | 11 12  9 10 |
|  3  4  1  2 |  7  8  5  6 | 11 12  9 10 | 15 16 13 14 |
+-------------+-------------+-------------+-------------+
| 14 13 16 15 | 10  9 12 11 |  6  5  8  7 |  2  1  4  3 |
| 10  9 12 11 | 14 13 16 15 |  2  1  4  3 |  6  5  8  7 |
|  6  5  8  7 |  2  1  4  3 | 14 13 16 15 | 10  9 12 11 |
|  2  1  4  3 |  6  5  8  7 | 10  9 12 11 | 14 13 16 15 |
+-------------+-------------+-------------+-------------+
| 13 14 15 16 |  9 10 11 12 |  5  6  7  8 |  1  2  3  4 |
|  9 10 11 12 | 13 14 15 16 |  1  2  3  4 |  5  6  7  8 |
|  5  6  7  8 |  1  2  3  4 | 13 14 15 16 |  9 10 11 12 |
|  1  2  3  4 |  5  6  7  8 |  9 10 11 12 | 13 14 15 16 |
+-------------+-------------+-------------+-------------+
```

**Answer** 9.3

One solution to minimize the number of trips is as follows. Let $t$ be time, or
the number of trips. At $t = 0$, there are three dogs and chicks on the left bank.
At $t = 1$, a dog and a chick cross the river to the right bank. At $t = 2$, the
chick returns to the left bank. At $t = 3$, two dogs cross the river to the right
bank. At $t = 4$, a dog returns to the left bank. At $t = 5$, two chicks cross the
river to the right bank. At $t = 6$, a dog and a chick return to the left bank. At

$t = 7$, two chicks cross the river to the right bank. At $t = 8$, a dog returns to the left bank. At $t = 9$, two dogs cross the river to the right bank. At $t = 10$, a dog returns to the left bank. At $t = 11$, the two dogs cross the river to the right bank, and all the dogs and chicks are on the right bank.

## Answer 9.4

One solution to minimize the number of trips is as follows. Let $t$ be time, or the number of trips. Let the three couples be $(A_h, A_w)$, $(B_h, B_w)$, and $(C_h, C_w)$. Subscripts $h$ and $w$ indicate a husband and a wife, respectively. At $t = 0$, there are three couples on the left bank. At $t = 1$, $A_h$ and $A_w$ cross the river to the right bank. At $t = 2$, leaving $A_w$ at the right bank, $A_h$ returns to the left bank. At $t = 3$, $B_w$ and $C_w$ cross the river to the right bank. At $t = 4$, $C_w$ returns to the left bank. At $t = 5$, $A_h$ and $B_h$ cross the river to the right bank. At $t = 6$, $A_h$ and $A_w$ return to the left bank. At $t = 7$, $A_h$ and $C_h$ cross the river to the right bank. At $t = 8$, $B_w$ returns to the left bank. At $t = 9$, $A_w$ and $C_w$ cross the river to the right bank. At $t = 10$, $B_h$ returns to the left bank. At $t = 11$, $B_h$ and $B_w$ cross the river to the right bank, and all the three couples are on the right bank.

## Answer 9.5

```
+----+----+----+----+----+----+
| 3  | 2  | 3  | 0  | 5  | 0  |
+----+----+----+----+----+----+
| 2  | 3  | 0  | 5  | 0  | 3  |
+----+----+----+----+----+----+
| 3  | 0  | 5  | 0  | 3  | 2  |
+----+----+----+----+----+----+
| 0  | 5  | 0  | 3  | 2  | 3  |
+----+----+----+----+----+----+
| 5  | 0  | 3  | 2  | 3  | 0  |
+----+----+----+----+----+----+
| 0  | 3  | 2  | 3  | 0  | 5  |
+----+----+----+----+----+----+
```

# Index

Milton Keynes UK
Ingram Content Group UK Ltd.
UKHW031132141024
449569UK00006B/246

9 781466 552630